Northern Illinois University

Decisions Against Nature: An Anthropological Study of Agriculture in Northern Thailand

Michael M. Calavan

Center for Southeast Asian Studies
Special Report No. 15, 1977

Exclusive Distribution By

THE CELLAR BOOK SHOP
18090 WYOMING
DETROIT, MICH. 48221
U.S.A.

PREFACE

The publication of Michael M. Calavan's monograph, based on research in the Chiang Mai area in northern Thailand, by the Center for Southeast Asian Studies is most appropriate for several reasons. First, Northern Illinois University, in cooperation with Chiang Mai University, has supported for four years a program, initiated by Professor M. Ladd Thomas, involving the exchange of faculty and graduate students of both universities. Second, the Center's program, since its establishment in 1963, has emphasized Thailand. Seven faculty members affiliated with the Center are Thai specialists, representing special interests in art history, history, language and literature, political science, and sociology. Third, the NIU campus has had, over many years, an able and stimulating community of young Thai scholars. It is hoped that this monograph, in addition to others published by the Center, will nourish the tradition of Thai Studies at Northern Illinois University.

Cultural ecology is a topic of growing importance to anthropologists. For example, a decade ago few, if any, introductory anthropology textbooks dealt with this topic. At present few such textbooks omit it. Ecology is concerned with the complex relationships among biological populations, the physico-chemical features of a given territory, and much concerned with processes of "adaptation." With the exception of humans, the adaptations of life forms to their habitat are severely limited by their genetic structure. Human capacities for adaptation are uniquely extended through the existence of culture. Hence, the cultural ecologist must also examine the "mutual dependency of the sexes, the nature and function of the nuclear family, the various extensions of such families, and still larger aggregates of families, bands, villages, and still higher levels of social integration such as tribes, states, and empires. He is also concerned with the various institutions in the areas of economics, religion and art" (*Anthropology Today* 1971:370).

Decisions Against Nature studies the adaptation of a community of Thai rice farmers to their locality by focusing on one major facet--crop choice. As stated in the first Chapter of this monograph: "Data presented are those that strike the author as most often relevant to crop choice. In developing the model, many facets of local behavior--agricultural ritual, irrigation societies, organization of labor--have been given short shrift." A mathematical model was then developed to test the value of this innovative approach to the understanding of ecological adjustment of Southeast Asian farmers.

Although Professor Calavan is modest in presenting his major conclusion, it is one that challenges basic concepts associated with agriculture and cultural innovation in developing nations. He found, contrary to other researchers, that Sansai farmers are "self-interested, flexible, economic maximizers, eagerly adopting innovations which promise economic advantage, and readily discarding crops, varieties, tools, and techniques which are no longer superior." The implications of this finding are of extraordinary importance both to the theoretical matrix of cultural ecology and current efforts to improve Asian agricultural development and rural living conditions.

Michael M. Calavan was born in Portland, Oregon. He graduated with a major in anthropology from the University of Oregon in 1965. In 1965 he began graduate work in anthropology at the University of Illinois at Urbana. The major focus of his studies, supported by NDEA fellowships, was human ecology. During the summer of 1967 he studied Thai at Yale University, and for most of the academic year 1967-68 nutrition, public health, agricultural economics, and Thai at the University of Minnesota.

During 1969-70 he did research in Sansai, supported by the Midwestern Universities Consortium for International Activities. He joined the faculty of the Department of Sociology-Anthropology at Illinois State University, Normal, in the fall of 1970. He received his Ph.D. degree from the University of Illinois in January, 1974. In 1975-76 he returned to Thailand on receiving a Southeast Asia Fellowship from The Ford Foundation. He revisited Sansai and did additional research in food use and nutrition. During both research periods in Thailand, the author was accompanied by Sharon Kay Calavan who received her doctorate in anthropology from the University of Illinois at Urbana in 1974. During the summer of 1977 Dr. Calavan traveled in South Asia to explore possibilities for future research in either India or Bangladesh.

The Center for Southeast Asian Studies

Donn V. Hart
Director

ACKNOWLEDGEMENTS

In preparing this monograph, I have incurred numerous personal and intellectual debts. I wish to thank my mentors at the University of Illinois at Urbana. Professor Clark Cunningham provided valuable guidance during the research period in Thailand and evaluated drafts of this study. Professor Demitri Shimkin gave me assistance in formulating my interests in human ecology, decision-making, and rice farmers. Professor F. K. Lehman taught me about Southeast Asia, and permanently aroused my intellectual interest in this cultural area. Professor Melvin Wagner and Professor Warren Peterson also were of great help in the organization of my materials. Miss Anong Supharatratana and Mr. Khongrith Thasanidamrongkun carried out their tasks as research assistants in Sansai with sensitivity and intelligence. The people of Sansai provided, with unfailing patience, instruction in matters of agriculture and general culture.

Department of Sociology-Anthropology
Illinois State University
Normal, Illinois

Michael M. Calavan

CONTENTS

Chapter One

INTRODUCTION

The northern Thai rice village of Sansai, located twenty kilometers southwest of Chiang Mai, is presented in this monograph as a population of decision-makers operating within a rule-setting environment. (See Map 1. Chiang Mai province is lined.) The village provides fertile

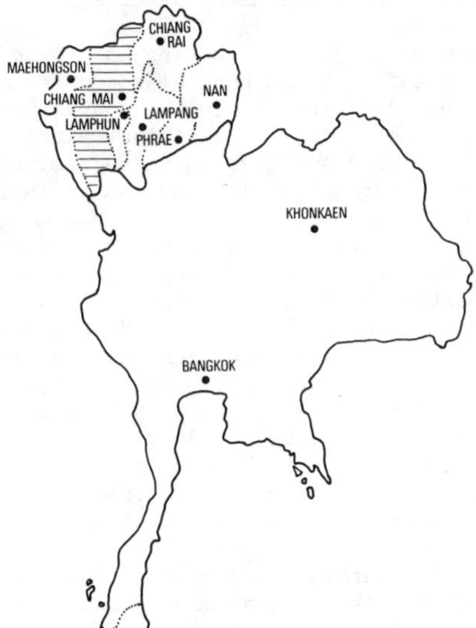

Map 1 *Thailand Excluding the Twelve Southernmost Provinces*

ground for construction of intellectual models of national and international significance. Farmers who raise wet rice with simple technology are found throughout Southeast Asia (and in other parts of Asia), Africa, Europe, and Latin American. It is believed that this model and its application to the problem of crop choice in Sansai suggests useful approaches to understanding agricultural adaptation in other parts of Southeast Asia and the world.

Sansai life is focused, in important ways, on activities related to rice production and consumption. This monograph describes some of these activities and analyzes the principles on which they are based. Analysis is based on data which are complex and quantitative--and sometimes inadequate. The approach used is an outgrowth of the writer's belief in the efficacy of quantitative data gathering and the great potential value of of mathematical decision models for the interpretation of human behavior.

Patterns of crop production at Sansai are examined in detail. Particular emphasis is given to the manner in which crucial decisions governing this agricultural activity are made. The model takes the form of a "decision against nature" in which it is assumed that farmers make difficult choices vis-a-vis a Nature which is unpredictable but not directly opposed to their interests. Basically, the model is inferred from quantitative data, and not derived from statements about choice made by the farmers themselves. However, farmers' opinions are not totally ignored. Their notions about significant local entities--land types, land use patterns, available crops and varieties, appropriate measuring devices--strongly affected data collection and analysis. Furthermore, the reader will find (especially in Chapter Four) direct quotations of strategic statements in which farmers reveal limited aspects of their choice procedures.

However, the "essence" of decision-making is assumed to lie beyond the range of these statements. Human choice procedures are *generally* so subtle and complex that decision-makers are unable adequately to recount them. This is probably true of decision-makers at all levels of education and in both developed and developing societies. It was decided to infer choice procedures through collection and analysis of quantitative data.

Data for this study were gathered with three concepts constantly in mind. First, that man is more usefully understood as being "in adaptation," and that any particular culture is best understood as a composite of "adaptive strategies". Second, that detailed studies of decision-making are capable of providing important insights into adaptive process. At Sansai, decisions involving crop choice were selected for study. Third, that quantitative measurement of inputs, outputs, and cyclical exchanges is necessary for adequate description of adaptive strategies and decision-making. Agricultural inputs quantified in this way include land, labor, fertilizer, and insecticide. Outputs are measured in terms of crop yields.

It is assumed that materials presented will be of greatest interest to the following three distinct audiences.

The first audience is Southeast Asianists. Descriptive materials on wet rice agriculture are likely to be of greatest interest. A point made previously by Hanks (1971), Moerman (1968), and King (1927) is re-emphasized in this monograph. Wet rice cultivation is a subtle and sometimes excruciatingly complex decision-making process. The impression that emerges from descriptive materials is of farmers who are progressive and open-minded, rather than conservative and tradition-bound.

The second audience is human ecologists. Several recent studies in ecological anthropology have stressed the *existence* of adaptation, while largely ignoring adaptive *process* (Rappaport, 1968; Meggers, 1971). The research described in this monograph represents an attempt to restore theoretical balance through study of strategic aspects of adaptation. The theoretical position implied by research results is that the intricate, highly adaptive pattern of Sansai agriculture has arisen largely through the concatenation of thousands of individually-taken, rational decisions.

The third audience is economic anthropologists and agricultural economists. The rationality (more specifically, the material self-interest) of at least one group of non-Western farmers is deliberately and carefully asserted on the basis of a model building and testing procedure. The notion of economic rationality is presented, accepted, and applied to development of a simple decision model. The model is then applied to a concrete choice situation in Sansai agriculture. It predicts successfully a sudden, major shift in varietal choice.

A Village in Adaptation

Data gathered at Sansai were concerned with such diverse topics as rice varieties, commodity dealing, labor exchange, and land prices. Obviously, some organizing concept is needed to raise these diverse observations above the level of simple description. The organizing concepts used are "adaptation" and "adaptive strategy." Data were collected on the basisof their assumed relevance to daily problems of adaptation to a constantly challenging environment.

Shimkin has argued convincingly for the application of mathematical decision models to problems in human ecology. He uses the key phrase "adaptive strategies," and defines them as "systems of cognition and behavior maximizing the well being and minimizing the hazards to survival of human populations" (Shimkin, 1966:38). The relationship between biological and decision factors in human nature is defined by Shimkin as:

> In all, man's biology--to wit: individuality, natural timing, physiological needs, interrelations with other species, information-based choices, and balance of genetic attributes and plasticities--sets an exacting framework of rules for decision-making in games of survival....The processes of human and other organic survival and evolution fit the requirements of a game as formulated by theory of Von Neumann and Morgenstern (1966:45).

This intellectual juxtaposition of human biology and mathematical decision theories promises rich insights into adaptive process. A series of significant issues--the value of pure versus mixed strategies, the role of "cheating", procedures for assignment of payoff functions, assignment of decision-making situations to positions along micro-mega and pure competition/pure cooperation scales--are raised, and ready-made approaches to model construction are provided. The major thrust of this monograph--development of a decision model applicable to crop choice at Sansai--is to render some of Shimkin's ideas into concrete, quantitative form.

Like most models, this one simplifies reality. It strips away many details of culture, behavior, and environment. Readers must decide for themselves whether the mathematical formulas developed represent a sterile exercise or approach the "essence" of agricultural choice in northern Thailand. The author prefers the latter evaluation. This seems to be a valid approach to explanation of agriculture as adaptive process. Data presented are those that strike the author as most often relevant to crop choice. In developing the model, many facets of local behavior--agricultural ritual, irrigation societies, organization of labor--have been given short shrift. Although these facets of behavior are clearly relevant to agricultural choice, broadly defined, they are not significantly related to specific crop choices examined in this study.

A major feature of Sansai's adaptive pattern is its dense population. The village has a population of 905 persons exploiting a land area of 1.35 km^2 (0.52 mi^2), thus attaining a density of 670 persons per km^2 (1720 persons per mi^2). Outside eastern Asia, rural population densities at this level would be considered unusually high. This high figure, and higher figures from the river valleys of Java, the Ganges Valley and Kerala Coast of India, the Pearl River Delta of China, and the Tonkin Delta of Vietnam, cry out for explanation. In recent years a number of scholars have attempted to provide at least partial explanation. Geertz has contributed the concept of "agricultural involution" and attempted to delineate the historical factors relevant to an understanding of the rapid growth and present density of Java's population (1963). Boserup and Dumond have speculated about the direction of causal links between food production technology on the one hand, and population growth and density on the other (Boserup 1965; Dumond 1965).

However, this monograph is not specifically directed toward evaluation of the hypotheses of Geertz, Boserup, or Dumond. Rather it is concerned with describing and explaining limited aspects of the long term processes which all three of these authors describe. The key question is: "Given the existence of an already dense population which is raising wet rice, what are the processes by which crucial decisions are made, and what do these tell us about wet rice agriculture as an 'adaptive strategy?'"

In particular, this monograph is concerned with description of a village in the Chiang Mai plain of northern Thailand. In a review of Moerman's *Agricultural Change and Peasant Choice in a Thai Village* Keyes has stated that:

> As a contribution to Thai studies, this volume is . . . to be welcomed for its excellent description of the type of 'frontier' community which has been so important in Thai history. It should be noted, however, that Ban Ping represents only one type of peasant community in Thailand. Too little attention has been paid, I believe, in the discussions of peasant social structure in Thailand to the differences between communities which still have contiguous lands available for new cultivation and those in which expansion of fields is no longer a possibility. It is to be hoped, therefore, that this pioneering sutdy will soon be matched by a study . . . of the cultural ecology of some Thai peasant community which has a 'closed' rice-based economy (1969:907).

Hopefully this monograph responds in part to Keyes' request. It is concerned with the cultural ecology of a "closed" rice village. However, at points this study differs from Moerman's book. There is no attempt at ethnosemantic analysis--a major thrust of Moermen's work. On the other hand, some effort is devoted to the development of mathematical models of crop choice, an area in which Moerman shows little interest.

In several senses, rice is a remarkable crop. It requires relatively high temperatures for growth, but it can be raised on a wide range of tropical and temperate soils. It is raised successfully in arid areas so long as the supply of irrigation water is ample. Its nutritional qualities combine with its ability to produce high yield on poor soils to make it one of the most "efficient" crops available. Only potatoes and a few root crops are capable of producing a higher yield of food energy on a given unit of land.

The paddy field is a highly efficient mechanism for protein production (*Oxford Economic Atlas of the World* 1965:26). With a little land, temperatures that do not fall below 18°C. during four or five months of the year, and an ample supply of water, there are few more efficient ways of producing edible protein than through rice cultivation. This explains why rice is the major food crop of tropical areas. These facts also partially explain the dense population which afflicts several regions of Asia. Such regions as the Tonkin Delta of north Vietnam or the river valleys of Java, could never have attained their present high population densities if any crop other than wet rice had been raised.

The problems arising from high population density at Sansai are shared with large portions of the Third World: shortage of land, small farm size, high land price,[1] underemployment, low wages, indebtedness, high interest rates, a high carbohydrate/low protein diet, and high levels of infant mortality. Other features of technologically simple, wet rice cultivation make a more positive contribution to rural life. For instance, a number of valuable supplementary foods are produced as by-products of rice production. Canals, irrigation ditches, and flooded fields produce several species of freshwater fish, snails, crabs, and aquatic plants. Fallow fields provide a number of wild vegetables. Pigs feed on the rice bran which is a by-product of milling. Rice straw provides cattle feed and mulch for garlic beds in Sansai.

The flooding of rice fields has important implications for rural life. Irrigation is a complex process which necessarily involves cooperation among many farmers for its construction and maintenance. Some type of irrigation association and a schedule of annual work days are inescapable "facts of life." Dams and major canals are often under central control, and bring rural people into close contact with the government. When Sansai rice farmers are visualized "in adaptation" to their major crop, the interrelatedness of several characteristics of that crop and their life styles becomes apparent. These ten features include:

(1) High potential of wet rice for protein and calorie production.

(2) A technology which involves use of simple tools, but requires construction and maintenance of complex irrigation systems and intimate knowledge of the physiological requirements of the rice plant.

(3) Shortage of land and excess of labor with high prices for the former and low prices for the latter.

(4) An annual social cycle which seems to ebb and flow with the labor requirements of the crop.

(5) Problems of too much water (resulting in death of rice plants by drowning) or too little (resulting in death of rice plants by desiccation) which are dependent variables, subject to variation in rainfall, decisions made by Irrigation Department officials, strength of protective dikes, etc.

(6) Commodity prices which are affected by local, national, and international demand.

(7) The availability of supplementary foods--fish, pork, chicken, ducks, gathered vegetables--which can be regarded as by-products of the cropping system.

(8) Nutritional qualities of the main crop which make it possible to subsist on a diet which derives ninety per cent or more of total calories from rice.

(9) Need for substantial human energy expenditures in raising the main crop and supplementary crops.

(10) Substantial socioeconomic variation, centered around control of land.

Decision-making at Sansai

Successful adaptation can be viewed as an ongoing process which involves making correct decisions. In recent years several anthropologists have emphasized the importance of decision models for understanding both diachronic and synchronic process. Robbins Burling has pointed out that:

> . . . we are 'economizing' in everything we do. . . . This economic view of society becomes one way, or if one prefers, one model for looking at society. It is a model which sees the individuals of a society busily engaged in maximizing their own satisfactions--desire for power, prestige, sex, food, independence, or whatever else they may be, in the context of the opportunities around them, including those offered by their own culture. Since one makes choices partly with an eye to the expected choices of others, it is not unreasonable to view this pursuit of satisfactions as a great and continuing *game of strategy*. . . .
>
> If we . . . focus upon the individual who is caught in the web of his society, and who is trying to maximize his satisfactions, we are led to the investigation of his *actual behavior in situations of choice* (1968:184 italics added).

Burling proposes that anthropologists study "actual behavior in situations of choice." However, he is not explicit concerning types of choice situations which are amenable to this form of observation. Neither is he explicit about particular mathematical models which may be useful in studying choice situations.

Raoul Naroll suggests the application of certain mathematical models to problems of cultural evolution. He calls his approach "stochastic epistemology" and speaks of human beings as "intuitive mathematical statisticians and Popperian logicians" (1970:27). He suggests that the ideas of a 17th century mathematician, Thomas Bayes, provide an adequate descriptive model of probability inference in cases of incomplete knowledge. Naroll summarizes Bayes' ideas thus:

> Whatever may be the degree of doubt or certainty about the most simple and seemingly obvious facts, there is no doubt in most people's minds about many simple but strong pleasant or unpleasant feelings. Most people know in large measure what they like and what they dislike. They must and in fact they do live their lives on the basis of their judgment--usually intuitive--of the probabilities, of the odds: first, the odds that their perceptions are accurate, or inaccurate; second, the odds that their theories about everyday living--most of which come to them via their culture, of course--are correct or incorrect (Naroll 1970:27).

Leonard Joy, a British economist who has studied agricultural decision-making among Indian farmers, offers the following insights into studies of this type:

> I have already made the point that primitive choice situations are not necessarily--nor even typically--simple. Typically, they are exceptionally complex. Almost all choice situations in cultivating communities that I have considered have been highly complex. By this I mean two things: first that there is a high degree of interdependence between activities such that small changes in cropping patterns, or the timing of operations, may have major ramifications through the whole farming system and affect related activities such as house-building, visits to the market, hunting, operations, ceremonials, and so on; secondly, that complexities arise because of the existence of uncertainties caused by variations in weather, incidence of pests, etc. It is the existence and extent of such complexities that requires the application of economic analysis (1967:40).

Joy provides fair warning that studies of agricultural decision-making lead to discussions of substantial complexity and subtlety. However, construction of models which clarify the principles underlying apparent chaos is the precise goal of anthropological (and economic) science.

In stressing the decision component of human behavior, Burling, Naroll, and Joy presumably assume the existence of a goal or set of goals toward which behavior is directed. Each author approaches the matter in a different way. Burling suggests the existence of a wide range of possible goals--power, prestige, food, independence, sex, etc. Naroll emphasizes the probability assumptions which precede choice and does not explicitly define the goals to which probabilities presumably are attached. Joy does not make explicit his assumptions about the nature of agricultural goals. However, it is reasonable to assume that, as an economist, he is inclined to emphasize material goals.

For the benefit of readers, the writer's assumptions about the nature of goals toward which agricultural choice at Sansai is oriented will be clearly stated: *It is assumed that Sansai farmers are motivated by material self-interest--i.e. by a felt need to provide themselves with food, clothing, housing, and other material items. It is further assumed that they regularly re-evaluate their farming operations in light of these goals.* Readers who distrust this assumption will find sufficient quantitative data to develop alternate models.

The data from Sansai are placed in broader perspective through examination of previous related work by Moerman, Behrman, Desai, Rogers and Shoemaker, Foster, Erasmus, and Davenport. Moerman's study of rice agriculture in a sparsely settled region of northern Thailand provides an interesting contrast with densely settled Sansai. Behrman uses aggregate data from many Thai provinces to arrive at conclusions not dissimilar to those of this study. Foster has discussed peasant decision-making in relationship to an inferred mental construct--the Image of Limited Good.

In his analyses of the behavior of Latin American farmers, Erasmus has emphasized the importance of "frequency interpretation and probable knowledge". Davenport has applied game theory to analysis of crucial choices made by Jamaican fishermen. Rogers and Shoemaker have examined the results of a large number of innovation studies and chosen to emphasize possible causal factors strikingly different from those considered in this monograph. The work of D. K. Desai is particularly illuminating. He applies linear programming techniques to evaluation of overall cropping patterns of forty Maharashtran farmers, and concludes that the patterns actually selected were not "optimum" (Desai, 1963). Desai's study will be examined in detail for two reasons. First, his conclusion regarding the "non-optimum" nature of agricultual decision-making in Maharashtra directly contradict the interpretation of Sansai decision-making. Second, his study demonstrates some of the intellectual complexities raised by serious attempts to explicate actual choice situations.

Measurement of Inputs and Outputs

Evaluation of particular decisions necessarily involves precise expression of alternate payoffs or utilities. For instance, the implications of a particular cropping strategy can be adequately understood only when quantitative data are available on inputs of land, labor, capital, and resulting yield.

Quantitative data on inputs and outputs were gathered for this study in a number of socioeconomic contexts. Information was gathered on the amount of land controlled by Sansai residents through mapping and surveys. A series of questionnaires provided data on labor and capital inputs, yield, timing, and variety of crop planted. Thirty households were interviewed concerning their annual spending patterns.

There are references to three separate socioeconomic groups within the village throughout this monograph. The distinction does not imply that sharply separated classes exist at the present time. Rather, its purpose is to demonstrate that there is substantial variation in economic resources and living standards within the village and to facilitate handling of statistical arguments.

The term "peasant" is consistently avoided. The cultivators at Sansai are called "farmers." The writer prefers to limit application of peasant to individuals who belong to a particular type of legally defined social group. No such definition can be applied at the present time to the residents of Sansai.

It seems appropriate to point out, in the beginning, that the model developed is inadequately tested. The model assumes that farmers use data on inputs and outputs from previous years in assessing probabilities and plotting strategies for the coming year. Because the field period was limited to fourteen months, input/output data are available for a single year. Therefore a completely adequate test of the model is not possible. However, it is believed that substantial evidence for the economic rationality of Sansai farmers is provided. If so, this is a far cry from final proof that *all* human behavior (or that of all Asian rice farmers) is based on rational decision-making. However, it is a modest step in the direction of defining if and/or where such rationality exists.

Chapter Two

RESEARCH METHODS

The data reported and analyzed in this monograph are drawn from observation, informal interviews, and seven formal research instruments. The purpose of this Chapter is to familiarize the reader with the formal research instruments and to provide him with a basis for evaluating their reliability and validity. The strengths and weaknesses of each instrument are discussed in detail. A general discussion of problems of reliability and validity then follows.

The Research Instruments

Research instruments discussed include: CUSSRI (Chulalongkorn University Social Science Research Institute) *Socio-Economic Profile Schedule;* Household Daily Schedules; Household Expenditure Survey; Late Rice Crop Survey; Garden Crop Survey; Early Rice Crop Survey; and Field/Crop Maps.

CUSSRI Socio-Economic Profile Schedule

In 1969-70, personnel of CUSSRI carried out a "Rural Labor Utilization Project" under a contract with the National Economic Development Board of Thailand. The *Socio-Economic Profile Schedule* was designed for the purpose of collecting socio-economic baseline data at research sites in Ayuthaya, Khon Kaen, and Chiang Mai provinces. The project director, Dr. Jacques Amyot, invited the writer and his wife to administer the *Socio-Economic Profile Schedule* at Sansai. The relationship was mutually beneficial. The CUSSRI project was initiated with an enlarged sample of baseline data. Moreover, the initial survey of population, landholding, agriculture, labor expenditure, and related activities at Sansai benefited from use of a carefully designed survey form. (See Appendix 1).

The *Socio-Economic Profile Schedule* was administered during October, 1969, with the assistance of eight students from Chiang Mai University. The survey was carried out more than three months after research was initiated at Sansai. Local residents had a reasonable amount of time to become acquainted with the researchers and their research procedures. Seven of the interviewers were native Northern Thai speakers. The eighth was a native of southern Thailand who had learned Northern Thai. The survey forms were printed in English and Central Thai, but all questions were asked, and answers received, in Northern Thai. Data were recorded in English, or where necessary in Central Thai. During the survey period, evenings were devoted to translation of difficult concepts, and checking the survey forms for internal consistency. Where necessary, interviewers were sent back to surveyed households to make corrections or collect additional data.

Table 12 (see Appendix 1) elicits information of great research interest, but it presented substantial problems in the Sansai context. Local patterns of double- and triple-cropping made it necessary to divide each section under item 1 into three portions. Instead of asking "How much farm land did you rent last year?", it was necessary for survey workers at Sansai to ask "How much land did you rent last year for early rice? Late rice? Garden Crops?" Frequently, the answers varied from one season to the next. Initially, recording these particular data was a source of confusion to the students, but great care was taken in checking and rechecking their answers at the time of the survey.

One household left the village a few months after the survey was completed. Another was missed during the initial survey, and the form was completed about three months later. At about the same time, a survey form was filled out with the assistance of a newly arrived household. Data from 186 households were included in aggregate village statistics, although there were never more than 185 households resident at any time during the research period. Data on agricultural patterns and labor expenditures cover the period from April 15, 1968 to April 14, 1969. The data refer to a period which ended a few months before field research was initiated. The acceptance rate was 100 percent. Every village household provided a respondent. No respondent openly refused to answer any specific question.

Household Daily Schedules

Ten households were selected to take part in a detailed survey of work patterns and annual cycle. One member of each selected household was asked to record daily work activities of all household members old enough to have finished primary school. (See Appendix 2). Representatives of all ten households agreed to take part in the survey. All recorders carried the survey through to the end. The enthusiasm and conscientiousness of recorders were bolstered through payment of a small stipend (25 *baht*) at the end of each month. (Twenty *baht* equals one American dollar). Data sheets were picked up twice each week, so that there was little opportunity for recorders to fall behind in their task. In one case, data were recorded by the household head, in another by the spouse of the household head. In all other cases, data were recorded by a younger member of the household, in his or her late teens or twenties.

Data collection diverged considerably from the "random" ideal. Households were selected about six weeks after initiation of research, and heavy emphasis was placed on selection of potentially cooperative households. The research assistants (both natives and residents of the village) were asked to select a cross-section of households--rich and poor, farmers and non-farmers. It was also necessary to select households with at least one literate member who was likely to be cooperative and reliable. In several ways, the selected sample corresponds to local population parameters rather poorly. Those few households with no literate members were excluded from the survey. Most of the selected households were close to the geographical center of the village and its road. Well-to-do households with non-agricultural workers are somewhat over-represented. In addition, several economically significant activities carried out by youngsters--fishing, baby-sitting, collecting edible insects, land crabs, snails, and "pig vegetables"--were not recorded.

Household Expenditure Survey

Data on annual expenditures were collected from thirty-one households. (See Appendix 3). The household head or his spouse was questioned once, and asked to remember or estimate expenditures for the entire previous year. These data probably are useless for comparing expenditures on *particular* items in *particular* households. However, it can be assumed that errors are random, making the data adequate for broader generalizations.

An economically stratified sample was drawn according to the following procedure. Data from the *Socio-Economic Profile Schedule* were used to construct a socioeconomic rating for each village household. Households were them listed in order from No. 1 to No. 186. Beginning with No. 3, every sixth household was asked to respond. A sample of thirty-one households was drawn. All selected households provided the information requested. Considerable time was devoted to the design of the survey form. The major concern was with construction of a nearly exhaustive list of potential expenditures.

Late Rice Crop Survey

A survey was carried out in December, 1969, for the purpose of collecting information on inputs of land, labor, and capital, on yields, and on timing of crucial operations for the early rice crop. (This survey form is nearly identical to the form used for the early crop. See Appendix 4). All village households who raise the crop within village boundaries were surveyed. Positive responses were received from all households in this category. Most of the survey forms were filled out in December, five months after initiation of research. A few were completed in February, after the delayed late rice harvest which sometimes occurs in late January and early February. Respondents were asked to provide information from one day to several weeks after harvest. The survey was administered by a research assistant, Mr. Khongrith, who is a native of the village and then taught in a nearby school. Occasionally he assists his father in cultivation of garlic. He also has a substantial familiarity with local agricultural practices.

The survey form was written in Central Thai, but all questions were asked in Northern Thai. Survey questions were formulated only after lengthy discussions with the research assistant, his father, his uncle, and several other residents. Particular attention was given to construction of an exhaustive list of agricultural tasks.

Garden Crop Survey

A survey of inputs and yields relevant to garden crop cultivation was done largely in March. Survey forms for a few households were completed after the peanut harvest in May. (See Appendix 5). The survey was administered by the same male research assistant who carried out the Early Rice

Crop Survey, and by a student from Chiang Mai University. Mr. Damrong, the second interviewer, served as an interviewer during administration of the *Socio-Economic Profile Schedule* in October. He is a native of a village north of Chiang Mai, in the Mae Rim District. He is quite familiar with local agricultural practices and his rapport with local residents was excellent. The survey was administered in Northern Thai, by native speakers, eight months after initiation of research.

All village households raising garden crops within village boundaries were surveyed. Respondents were questioned a few days to a few weeks after harvest. Several local residents were consulted during preparation of the survey form. Particular care was devoted to construction of exhaustive lists of agricultural operations and production expenses potentially associated with each crop.

Early Rice Crop Survey

All Sansai households raising early rice within the village were surveyed during late July, August, and early September of 1970. (See Appendix 4). A few forms were completed by the male research assistant after the research period had ended. The research assistant was assisted in this task by his younger brother, a student on vacation from the Teacher's Training College at Chiang Mai. Both survey workers were village natives, and had themselves worked in the fields. The survey form was written in Central Thai, but all questions were put in Northern Thai. Respondents provided information on expenditures of land, labor, and capital, and on yields, a few days to a few weeks after harvest.

Field/Crop Maps

Three Chiang Mai University students were given the task of constructing a village map. They used a plane table to measure the boundaries of every house compound and agricultural plot. They were accompanied and assisted throughout this task by two village residents, the village headman and a former assistant headman. Both were experienced tax collectors, and highly familiar with local field boundaries. A map, showing the location of all individually owned agricultural plots and house compounds, was constructed at a scale of 1:5000.

Copies of the village map were used to record data on crop choice over a period of three years. Data were collected during July and August, 1970, by a young man who is a native of the village, but has received ten years of education in Sansai and Chiang Mai. During the two years following his return to the village, he often had worked as an agricultural laborer, and thus possessed a fair knowledge of owners, renters, and field boundaries.

He approached each household head, and asked questions of the form: "Which rice variety did you choose for the early crop in 1970? 1969? 1968?" Equivalent questions were asked for each of the other crop seasons. Where a single plot was divided between two or more rice varieties (fairly common for late rice), or two or more garden crops, it was necessary to estimate the total area devoted to each. The interviewer discussed the matter with the respondent for a few minutes, and then recorded his estimate of relative proportions in a rough sketch. The data collected are presented as maps in Chapters Six and Seven.

Problems of Reliability

When social scientists reach the point of testing specific hypotheses, they must concern themselves with problems of reliability. For example, they must be concerned with the reproducibility of data which support or contradict a particular hypothesis. Gould and Kolb define the concept of reliability as follows:

> *Reliability* is used by social scientists in the sense of degree of stability or reproducibility of empirical results. Science is based on the assumption that its findings are not unique but can be duplicated under identical conditions. Experience, however, has shown that identity is an ideal which can only be approximated in the empirical world. No matter how controlled the conditions, measurements vary. This variation may be due to systematic influences (non-random or constant errors) or to non-systematic influences (random or variable errors). The latter are the ones of importance in a determination of reliability, although both are of concern to the validity of the results (1964:587).

Data on agriculture and crop choice at Sansai are reliable to the extent that another researcher could have gone into the same situation at the same time, and derived essentially the same figures for labor and capital expenditures, yields, prices and other variables related to economic decision-making. Since this approach is subject to temporal restrictions--the period and special conditions to which the data are tied are past and not easily reproduced--some other approach to the evaluation of reliability is required.

Psychologists often are faced with this problem. It is impossible, inappropriate, or meaningless to repeat many psychological measurements. The researcher must apply some test of internal consistency which demonstrates that the magnitude of random error has been successfully held in check. There are several statistical procedures which yield a "reliability coefficient" as an index of reliability. The reliability coefficient has been defined as:

> A coefficient introduced by Spearman (1910) into psychology. Its object is to assess the systematic component of a variate (test) as distinct from error components. In psychology it is usually measured by the correlation between the result of two administrations of the same test. The 'reliability' as a quantity is the complement of the error variance of the test. . . (Kendall and Buckland, 1971:129).

A high positive correlation between two data sets indicates a small error variance, and thus "high reliability." In cases where it is impossible to collect repeated data sets, it is necessary to adopt some other approach to assessment of reliability.

A rather simple approach was adopted by the writer. If a sample is "reliable" or "reproducible," it should be possible to draw random sub-samples with parameters similar to each other, and to the total sample. For example, the mean of each sub-sample should conform closely to that of each other sub-sample, and to that of the total sample. The same is true of other parameters. Thus, the degree of convergence between sub-sample parameters becomes an indicator of reliability. Some of the data from Sansai were subjected to the following tests. Following a table of random members, yield data from the early and late rice crops, and from the garlic and melon crops were divided into sub-samples of equal or near-equal size. The sub-samples were subjected to the following tests:

$$\bar{X}_a \approx \bar{X}_b \approx \bar{X}_{ab}$$

Where $\bar{X}_a$ = arithmetic mean of sub-sample a

$\bar{X}_b$ = arithmetic mean of sub-sample b

$\bar{X}_{ab}$ = arithmetic mean of the total sample

$$S_a^2 \approx S_b^2 \approx S_{ab}^2$$

Where S^2 = variance of a sample

$$\delta_a^2 + \delta_b^2 = \delta_{ab}$$

Where $\delta^2 = X - \bar{X}$

Table 1 presents data on sub-samples of yield data drawn at Sansai. A high degree of convergence is indicated in all four cases.

Problems of Validity

In addition to a concern with the size of random error, the hypothesis-testing social scientist must also be concerned with problems of *validity*. Although it has been established that random error has been held to a reasonable level, it is necessary to demonstrate that the data are truly relevant to the hypothesis being tested. The data collection techniques also must be relatively free of non-random error.

> *Validity* and its cognates (valid, validate, validation) are used by social scientists in three related senses: denoting (a) soundness of conceptualizing, (b) applicability of research techniques, and (c) pertinence of data. In all three cases the emphasis is on the value or worth of the idea, the technique, or the datum for some specified objective, as judged in relation to some standard or criterion (Gould and Kolb, 1964:742).

It is quite possible for data to be reproducible (reliable) and not valid. They may be unrelated to the hypothesis being tested. They may also result from repeated application of a research technique which skews reality in a consistent manner. For example, a scale which consistently weighs items one kilogram too heavy will provide data which are reliable, but not necessarily valid. Accurate measurements applied to some population may be reliable, but not valid because they are irrelevant to the hypothesis being tested.

Table 1 *Indicators of the Reliability of Research Instruments Used at Sansai*

		Sub-sample a	Sub-sample b	Total Sample
Early Rice Yield	N =	29	28	57
	$\overline{X}$ =	46.1	51.7	48.9
	s^2 =	218.1	242.5	233.4
	$\Sigma\delta^2$ =	6,107	6,548	13,070
Late Rice Yield	N =	45	44	89
	$\overline{X}$ =	27.2	29.9	28.6
	s^2 =	195.5	226.0	206.2
	$\Sigma\delta^2$ =	8,602	9,718	18,148
Melon Yield	N =	28	28	56
	$\overline{X}$ =	798	747	772
	s^2 =	157,751	176,849	165,707
	$\Sigma\delta^2$ =	4,259,277	4,774,523	8,782,471
Garlic Yield	N =	42	41	83
	$\overline{X}$ =	790	805	796
	s^2 =	194,656	171,555	184,954
	$\Sigma\delta^2$ =	7,980,896	6,862,200	15,134,237

While problems of reliability pose an important first hurdle, assessment of validity is the major problem in hypothesis testing. Has the researcher demonstrated what he set out to demonstrate? Was it worth doing? Did he use appropriate techniques for data collection and analysis? In assessing the validity of his research, the researcher must anticipate criticism, admit existing weaknesses, and clarify areas of strength. The strengths and weaknesses of the model developed in this study can probably be assessed best by posing, and answering, a set of questions. The questions parallel Gould and Kolb's three "senses" of validity: First, does the model adequately encompass the complexities of economic "rationality?" Second, were data collected with sufficient care to insure accurate measurement of inputs and yield? Finally, is collection of data on inputs, yields, and crop choice really relevant to exposure of the "rationality" or "irrationality" of Sansai residents?

Soundness of Conceptualizing

It is clear that our model is simpler than reality. It equates the complex patterns of information gathering with a simple random process. It assumes that all individuals use the same basic techniques for evaluation and choice. Alternate strategies are evaluated in terms of "yield," a scale which seems excessively simple for most scientific purposes. To these anticipated criticisms, two rejoinders seem appropriate. First, the model seems to "work" and second, lessens the complexities of model construction for a relatively simple choice situation was examined.

In a modest way, the model appears to work. It predicts a major shift in varietal choice for the early rice crop between 1970 and 1971. Although its precise dimensions are not known, it is clear that such a shift did take place. The model predicts a range of possible outcomes, and several parametric variations were "plugged in" to the model. But all possible outcomes predicted that a substantial majority of Sansai households would plant *khaaw kaseet* in 1971. The incomplete data available suggest that the true figure was about 70-90 percent. One possible reply to the criticism that the model seems "too simple" is that "It works."

The model *can* be relatively simple, because a relatively simple situation was selected for examination. Choosing between *varieties* (e.g. the early rice crop at Sansai) is essentially a simpler process than choosing between *crops* (e.g. the garden crop at Sansai). For a given planted area, all rice varieties require essentially the same input of labor and capital. This is not true for the garden crops raised at Sansai. Mean labor and capital requirements of the four crops vary greatly. For the garden crop, simple measurement of yield is clearly inadequate. Sansai farmers do not apply mixed strategies in choosing a variety for the early rice crop, although it is commonplace to do so for the other crops. It seems likely that mixed strategies are not applied to the early rice crop because they are all inferior to at least one pure strategy. Since the possibility of mixed strategies was ignored, the model was greatly simplified. Another possible reply to the criticism that the model is "too simple" is that "It should be simple, because it has been deliberately applied to a relatively simple situation."

Applicability of Research Techniques

The model developed in this research is testable in quantitative terms. Adequate testing clearly depends on adequate techniques of data collection. The bulk of the data presented in this monograph were collected in a series of surveys of modest scope. Each of the surveys has been described. Considerable care was taken to limit both random and non-random errors in data collection. The following are "strong points" of the data collection process.

Survey forms were written only after a series of interviews with local residents. In each case, great care was taken in formulating questions of genuine relevance to the topic at hand.

All face-to-face research was conducted in Northern Thai. Although many residents are fluent in Central Thai, Northern Thai is the language of daily discourse. Except for the writer, his wife, and one student from Chiang Mai University, all individuals involved in data collection were native speakers of the Chiang Mai dialect of Northern Thai. It seems unlikely that language problems were a significant source of random or nonrandom error in the data collection process.

In general, residents of Sansai were not asked to respond to formal research instruments until some months after initiation of research. During the intervening period they had substantial opportunity to become acquainted with the writer and his wife. We resided in the village an average of six days each week. Residents became accustomed to seeing us in the morning market, at the temple, attending weddings and funerals, visiting homes and fields.

Our association was with Sansai and its residents, not with government officials. We associated with the latter only on a few, formal occasions. Our permanent research assistants were natives of the village who were able to reassure residents regarding the harmlessness of our intentions. Perhaps the best indication of rapport with local residents is that all research instruments described in this chapter received 100 percent response. All households that were asked to provide information (most were asked three to seven times) did so. It is unlikely that random or non-random errors of data collection resulted from lack of rapport.

Of data collected, only those in the *Socio-Economic Profile Schedule* were made available to officials of the Thai government. The latter data, punched on computer cards, are in the hands of CUSSRI. The survey forms, and the names of respondents, are in the hands of the writer. The anonymity of Sansai residents have been substantially protected. They did not have to worry about government administrative actions in response to their answers.

Estimates of cropped area and production derived from survey materials receive modest independent confirmation from estimates of food intake. Total production of late rice (glutinous varieties) in the period from August, 1969 to July, 1970 was estimated at 154 tons. This coincides closely with an estimated annual rice requirement for the village of 173 tons (Calavan 1973:23-24). If we allow for modest use of non-glutinous varieties, estimated values suggest that Sansai is self-sufficient in glutinous rice. This conclusion fits well with observations that little glutinous rice is exchanged between villagers and the outside.

Examination of population data from Chiang Mai Province, Saraphi District, and Sansai provides another source supporting the validity of data gathered at Sansai. All three data sets show that age groups born in the early and middle 1940's and middle and late 1910's are smaller than adjoining age groups. No doubt these figures reflect the world-wide influenza epidemic which came at the end of World War I, and the smallpox epidemic which struck northern Thailand following World War II.

Pertinence of the Data

Crop choice at Sansai conforms to established definitions of "economizing" or "maximizing" behavior. The basic conditions of economic choice are present:

> The ends are various. The time and means for achieving these ends are limited and capable of alternative application. At the same time the ends have different importance (Robbins, 1968:94).

Sansai residents choose among varieties or crops with different potential yields and profits. They operate with limited resources of land, labor, and capital. They are well aware that they face a variety of choices with a variety of possible outcomes. Clearly, they face situations of economic choice. For these reasons the data collected appear to be pertinent to the problem at hand. If Sansai behavior conforms to a model based on concepts of "rationality," there is a basis for calling that behavior "rational."

Chapter Three

THE VILLAGE: SETTING, POPULATION, AND ANNUAL CYCLE

The research for this study was carried out at Sansai between July, 1969, and August, 1970. The physical setting within which daily life and agriculture occurs are briefly described. Statistical insights into the local population are then provided. Finally data collected with daily work patterns in ten households are used to describe the ebb and flow of life at Sansai.

The village lies roughly twenty kilometers southwest of Chiang Mai on an unpaved road which branches off the Chiang Mai-Lamphun highway. (See Map 2. The black squares indicate provincial capitals, the black circles district centers.) Sansai is located in the southern portion of Saraphi District, about one kilometer north of the provincial boundary with Lamphun.

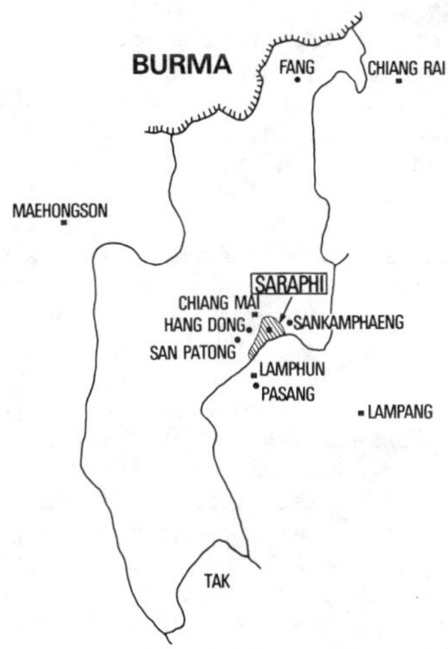

Map 2 *Chiang Mai Province*

Sansai is near the center of a Subdistrict (*tambon*) of the same name. Sansai Subdistrict is bounded on the west by the Ping River, on the south and east by Lamphun Province, and on the north by Khua Mung and Thaa Kwaang Subdistricts. The village lies about seven kilometers southwest of Ku Daeng (Kingshill 1965) and about the same distance from "Chiangmai" (pseudonym for village described in Potter 1976).

The Setting

Saraphi District lies near the center of the Chiang Mai plain, the largest of northern Thailand's upland valleys. Saraphi is flat, has no extensive forests, and is intensively cultivated. It has a population in excess of 60,000 and a total area of about 125 km^2 (Cunningham and Doege 1969:3). Thus the District's population is about 480 persons per km^2. Residents adapt to high population density through intensive cultivation of rice and garden crops. Much agricultural land is cropped two or three times each year. A complex irrigation system makes this possible.

Although Sansai is primarily an agricultural village, many residents devote at least part of their time to a variety of commercial and craft activities. Included in the former category are: market women, shopkeepers, and middlemen dealing in rice, longans (a local fruit), garlic, peanuts, and soybeans. Included in the latter category are: seamstresses, sawyers, carpenters, mat-weavers, basket-makers, and a variety of medical practitioners.

Residents live in expensive wooden houses, or inexpensive bamboo houses They purchase food and other items at two daily markets, and several "general" shops. Daily truck service provides transportation to Chiang Mai. The village has a school of four grades and a Buddhist temple (*wat*). Residents regularly speak of three major "problems" within the village: lack of an intermediate school; the poor condition of the village road; and the lack of electric power lines. Although small, Sansai is a socioeconomic unit of substantial complexity.

Map 3 provides a closeup view of Sansai and its immediate surroundings. Including the fields which are taxed as part of its administrative area, the village has a maximum length of roughly 1.9 kilometers and a maximum width of approximately 1.1 kilometers. The area of the village is close to 1.35 km^2. The approximate area of living space is 0.39 km^2. On this area are located: 185 houses, two rice mills, a noodle shop, several general shops, the morning and evening markets, a Buddhist temple, a school of four grades, the village spirit houses, roads, paths, canals and irrigation ditches. There are also several longan orchards, kitchen gardens, pig pens, other structures, trees, plants, and animals that are typically part of the residential area of a northern Thai village.

The village residential area lies slightly higher than the fields which surround it. Within the 1.35 km^2 which constitute the total village area, altitude varies by less than 1.5 meters. However, such minor variation is significant. During dry spells a variation of only 25 centimeters may spell the difference between inadequate and plentiful irrigation water. During periods of flood it may mean the difference between a successful rice crop, and one which is affected severely by drowning.

Depending upon altitude, availability of irrigation water, and other factors, Sansai's fields may be used for one, two, or three annual crops. Most fields are used for two or three. Each of the annual cropping seasons

Map 3 *Sansai, Settled Area*

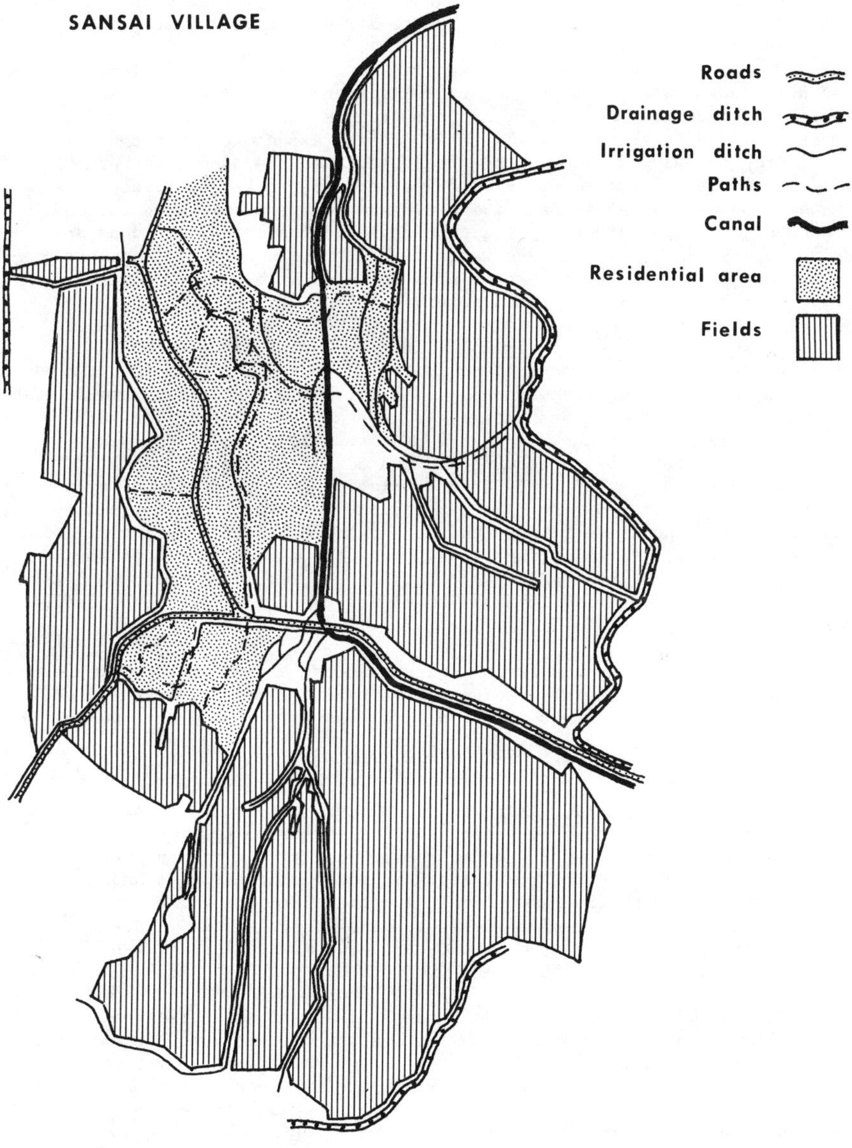

is given a name by the villagers. *Naa dau* is the "early rice" crop. Seeds are ordinarily sown in the nursery in March, the seedlings transplanted during April, and the crop harvested in July. The main problems associated with this crop are inadequate irrigation water early in the season, and rising flood waters at harvest time.

Naa pii is the second or "late rice" crop. Seeds are ordinarily sown in the nursery in late June or July, transplanting takes place between late July and early September, and the crop is harvested in late November, December, and later months. The main problem associated with the late rice crop is heavy flooding soon after seedlings are transplanted.

Residents speak of the third annual crop as "making a garden" *(tham suan)*. Crops planted at Sansai include garlic, peanuts, soybeans, and watermelons. Depending upon the crop chosen, planting may be done any time between late November and February. Harvest occurs between late February and mid-May. The main problem associated with this crop is drastic fluctuations in off-farm prices received.

During the spring and early summer of 1968, ninety Sansai households raised 316.75 *rai* of early rice. (One *rai* equals 0.16 hectare.) During the late summer and fall of 1968, 120 households raised 493.25 *rai* of late rice. During the winter of 1968-69, 136 households raised 324.50 *rai* of garden crops. Of this area, 191.50 *rai* were in garlic, 53.25 in soybeans, 49.25 in peanuts, and 24.0 in watermelons. Nine *rai* were planted in minor crop--tobacco, corn, and out-of-season rice.

The soils of Sansai's fields are alluvial. A map published by the Soil Survey Division of the Land Development Department of Thailand types the soils of land lying between Chiang Mai and Lamphun as "alluvial soils, on recent fresh-water alluvium; level to gently undulating (Moorman and Rojanasoonthon 1967). The geographer Robert Pendleton included Sansai in the area of "Chiangmai Loams." He offers the following two descriptions of this soil type:

> These are recent alluvial soils and some stream terraces in the mountain valleys. These soils are best developed in the Chiangmai Valley, where the Ping and tributary rivers have deposited fine sandy loams, silt loams, and clay loams in the lower portions of the valley. Here they are used for the main summer padi crop. Under irrigation, peanuts, soybeans, and near Lamphun, garlic, are grown in winter. In the lower portions of the terrain where the soil is apt to be too wet for soybeans or peanuts, a second crop of padi may be raised. . .
>
> Experiments on Chiangmai loams indicate that nitrogen and phosphorus fertilizers seldom, if ever, appreciably raise crop yields. However, residual effects on a following crop, particularly a legume, may make the use of commercial fertilizers worthwhile (1962:80-81).

The fields lying within the borders of Sansai are irrigated by a single large canal and the numerous irrigation ditches which lead from it. The system is gravity-fed, and the flow of water is generally to the south. The name of the canal is *Myang Caaw Mahawong*, in honor of the Chiang Mai prince who moved to Sansai in the latter part of the nineteenth century and supervised construction of this and two other canals. It flows eleven kilometers from Don Kaew Subdistrict, through Khua Mung Subdistrict, and through Sansai. At its head it is controlled by a water gate which allows water to flow out of another, larger canal. The larger canal receives water from another canal which flows directly from the Ping River.

Water flow in the *Caaw Mahawong* canal is controlled by three gates, all under the control of the Royal Irrigation Department of Thailand. At its foot, the canal changes form, and becomes a drainage ditch. Provided that the Ping is not too high, excess water can be drained back into the river through another water gate. This gate is controlled by local irrigation headmen. Throughout most of its length, the canal is about two meters wide. The irrigation ditches that branch off the canal are between forty and eighty centimeters in width. During the early rice and the garden crop seasons, irrigation headmen must sometimes visit officials of the irrigation station at Don Kaew and request that the water gates be opened. The flow of water onto individual fields is effected by construction of temporary earth dams across irrigation ditches, and by temporarily breaking the bunds that surround the fields.

Irrigation is a complex business involving competition for scarce water resources. Much local land is used for three annual crops, calling for delicate decisions in the use of water. Farmers who use water from a particular canal elect one of their number as irrigation headman (*hua naa myangfaaj*), with authority to settle disputes and oversee maintenance operations. All owner-cultivators who use water from a particular canal are obligated to assist each year in cleaning and improving it. This obligation is known as *kaakaan*. In cases where land is rented, owner and renter bargain about which will accept this obligation. Wealthy landowners who have this obligation hire a substitute to do the actual work.

Weather data were collected in the village from the beginning of August, 1969, to the end of August, 1970. Table 2 summarizes data on minimum, maximum, and mean temperatures on a monthly basis. Data on Chiang Mai temperatures are derived from *World Weather Records, 1951-60*. The record shows that overnight lows remain in the medium 20's Centigrade from April to November. Between November and March overnight lows are below 20°C., and in December, January and February, temperatures sometimes fall below 15°C. The lowest temperature recorded during the research period was 10°C. in December, 1969, and January, 1970. Between November 28, 1969, and March 11, 1970, there were 39 nights when the temperature fell to 15°C. or lower.

Maximum temperatures tend to be highest in March and April, before the start of the southwest monsoon. Mean temperatures recorded during one year in Sansai are consistently higher than those recorded over a ten year period at Chiang Mai Airport. There are several possible explanations for this. First, 1969/70 may have been an exceptionally hot year. Second, the Chiang Mai Airport may have slightly cooler weather because it is somewhat higher than Sansai. This airport is located on the lower slopes of a mountain, Dauj Sutheb. Third, the minimum/maximum thermometer may have been set up incorrectly. It is difficult to find a place that is protected from direct sunlight throughout the year. A correct explanation would probably incorporate elements of the last two possibilities.

Table 3 summarizes data on rainfall in Sansai during a thirteen month period, and on rainfall patterns in Chiang Mai during a ten year period. At first glance, both Sansai and Chiang Mai seem to receive substantial amounts of rainfall. The annual average for Chiang Mai for the 1951-60 period was 1323 mm. From August, 1969, to July 1970, Sansai received 1445 mm. of rain. However, average or mean figures are a poor guide to the

Table 2 *Monthly Mean Temperatures, Chiang Mai and Sansai*

	Aug.	Sept.	Oct.	Nov.	Dec.	Jan.	Feb.	Mar.	Apr.	May	June	July	Aug.
Sansai - Mean Daily High Temperature, by Month (Aug. 69 - Aug. 70)	33.20	35.7	37.2	34.4	33.0	35.1	37.8	39.8	40.4	36.8	35.7	34.4	35.4
Sansai - Mean Daily Low Temperature, by Month (Aug. 69 - Aug. 70)	23.5	23.5	22.6	19.8	16.2	16.4	14.8	17.2	22.0	23.1	24.4	23.4	22.7
Sansai - Mean Daily Temperature, by Month (Aug. 69 - Aug. 70)	28.4	29.6	29.9	27.1	24.6	25.8	26.3	28.5	31.2	30.0	30.1	28.9	29.1
Chiang Mai - Mean Daily Temperature (Aug. 69 - Aug. 70)	27.0	26.8	26.2	24.4	21.4	21.3	23.1	26.0	29.0	28.8	27.9	27.4	27.0

Table 3 *Patterns of Rainfall, Chiang Mai and Sansai*

	Aug.	Sep.	Oct.	Nov.	Dec.	Jan.	Feb.	Mar.	Apr.	May	June	July	Aug.
Monthly Total - Sansai, (Aug. 69 - Aug. 70)	230mm.	128	144	25	11	Tr	0	56	116	299	209	128	337
Monthly Mean Total - Chiang Mai, (1951-60)	215	284	133	29	11	17	13	24	54	180	178	188	125
Minimum for Month - 10 Year Period	163	208	20	Tr	0	0	0	0	6	66	39	54	163
Maximum for Month - 10 Year Period	302	474	291	83	85	59	89	68	123	326	347	260	302
Standard Deviation	73	76	79	29	26	21	27	26	44	74	88	61	73

adequacy of rainfall for agricultural purposes. Without some knowledge of variance, such figures are nearly worthless. During the period 1951-60, the reporting station at Chiang Mai Airport experienced substantial variation in rainfall. During 1953 the total was 2032 mm. During 1959 it was only 958 mm., less than half as much. One must evaluate the rainfall of the Chiang Mai Plain as unreliable, although average figures are moderately high.

Monthly standard deviation figures in Table 3 demonstrate that potential variation is great in all months of the year. Even the "dry" months of December through March may receive 50-75 mm. of rain in some years. But variation is greatest during the monsoon rains and greatest of all during the months in which they begin (May and June) and end (September and October). In conjunction with a "mean" rainfall of 178 mm., the standard deviation for June of 88 mm. can be interpreted in the following manner: "It is reasonable to expect that during one June in three, rainfall levels will fall below 89 mm. or rise above 266 mm." Rainfall data for the crucial month of April suggest the following interpretation: "During one year in three, April rainfall levels will fall below 9 mm. or rise above 98 mm." Almost certainly, similar variability is characteristic of rainfall at Sansai and is cause for substantial doubt and anxiety among farmers.

Population Profiles

Population characteristics and economic patterns are interrelated in significant ways. Allusion has already been made to the possible effect of Sansai's dense population on agricultural practice. Also significant in interpretation of economic activities are age, sex, and socioeconomic profiles.

Ethnically and religiously, Sansai residents are a homogeneous group. With three exceptions they are natives of northern Thailand. One exception is a Chinese tailor who resides in the village only occasionally, and is a native of Kwangtung province in China. A second is another man of Chinese ancestry who operates Sansai's largest general shop; he was born in Phitsanulok province. The third non-northerner is a school teacher who was born in Nakhorn Ratsima province. All others consider themselves to be "native people" (*khon myang*) of northern Thailand. However, the homogeneity of Sansai's population ought not to be overstressed. Several villagers claim Shan or hill tribe ancestry. Several were born in Thai Lue villages in Lamphun; their native dialect is quite different from the Chiang Mai standard. With a single exception, villagers were raised as Buddhists. The exceptional individual is a native of the nearby Moslem village who married a Sansai woman.

Sansai's population consists of 905 individuals who live in 186 ordinary households, and six or eight men and boys who ordinarily reside at the village temple. Due to births, deaths, and migration, village population varied slightly during the year. Average household size is 4.87 persons.

Table 4 represents Sansai's population when surveyed in October, 1969. A number of patterns which are normal and expected in "developing" societies are present: (1) the number of males and females are nearly equal; (2) children (younger than 20) outnumber adults in the childbearing age groups (roughly 20 to 50); and (3) generally, the older age groups are smaller than the younger. But certain anomalies exist in age group size and sex/age distributions.[1]

Table 4 *Population of Sansai, 1969*

Age Group (Years)	Number of Males	(Percent of Total)	Number of Females	(Percent of Total)
0 - 4	52	(5.7)	54	(6.0)
5 - 9	68	(7.5)	71	(7.8)
10 - 14	84	(9.3)	70	(7.7)
15 - 19	50	(5.5)	38	(4.2)
20 - 24	20	(2.2)	19	(2.1)
25 - 29	16	(1.7)	16	(1.7)
30 - 34	28	(3.1)	36	(4.0)
35 - 39	35	(3.9)	28	(3.1)
40 - 44	23	(2.5)	28	(3.1)
45 - 49	21	(2.3)	29	(3.2)
50 - 54	12	(1.3)	17	(1.9)
55 - 59	13	(1.4)	4	(0.4)
60 - 64	11	(1.2)	13	(1.4)
65 - 69	9	(1.0)	13	(1.4)
70 - 74	3	(0.3)	9	(1.0)
75+	6	(0.6)	9	(1.0)
Totals	451	(49.5)	454	(50.0)

Data from Sansai suggest that the birth rate is falling off. (Each of the two youngest age groups is smaller than the one preceding it.) In part, this can be associated with the introduction of birth control pills into the area. However, of greater probable importance is the small size of the 20-29 age group. Under ordinary circumstances, the majority of babies will be born to mothers in their 20's. Ten years from now, Sansai's birth rate may well be on the rise again.

The "dependency load" of a given population can be defined as "that portion of the total population too old to do significant physical work (65 years or older) and that portion of the total population too young to do significant physical work (0-14 years)." Table 5 indicates that Sansai"s dependency load is unusually high. In areas where land is plentiful and labor is in short supply, this would be a matter of some concern, i.e., a chronic labor shortage would exist. However, at Sansai no such shortage exists. The large portion of the local population aged 0-14 years indicates continued, growing pressure on already strained land resources.

Table 5 *Dependency Loads in Sansai and other Political Units*

Unit	Year	Dependents 0 - 14	65+(%)	Workers 15 - 64(%)
Sansai	1969	44.1	5.4	50.5
Saraphi district	1970	41.5	4.7	53.8
Chiang Mai (*myang*) district	1970	39.1	3.6	57.3
Chiang Mai province	1970	41.2	3.5	55.3
Thailand	1970	43.1	2.8	54.1
India	1970	37.0	6.0	57.0
Japan	1970	30.0	9.0	61.0
United States	1970	31.0	13.0	56.0

An obvious "spurt" in Sansai's population after 1950 must be attributed to the success of a malaria control program initiated in the area by the World Health Organization in 1948-49. The result has been a reduction in incidence rate from 40 or 50 percent in 1949, to three reported cases in Saraphi in 1969 (Cunningham and Doege 1969:3). Because malaria makes heaviest inroads into the youngest portions of the population, the effect of malaria eradication has almost certainly been to lower infant and child mortality rates, and to lead directly to rapid population growth.

The population of Sansai and surrounding villages continues to grow at a substantial rate. Unfortunately, local records of births, deaths, and migration are not sufficient to provide accurate figures for growth rate. However, the population of Saraphi District grew from about 40,000 in 1952 to roughly 60,000 in 1969 (Cunningham and Doege 1969:3). This represents an annual growth rate on the order of 2.0 to 2.2 percent. It is not unreasonable to assume that population growth in Sansai Subdistrict and Sansai has proceeded at a similar rate.

Table 6 contains data on population densities in Saraphi District, Sansai Subdistrict, and Sansai village. In each case, population density figures represent the relationship between local population and the land area of a political unit. But the amount of land owned and worked by residents of any one village does not necessarily coincide with the amount of farm land within the boundaries of the village. Villagers work some land outside the "home" village, outsiders work some land within it. Sansai residents actually own 623 *rai* of field land and thus attain a "nutritional density" of 924/km^2. Figures for individual villages must be examined with caution. However, the figures for the subdistrict and district levels may be accepted with some confidence. Population density in the immediate area of Sansai is on the order of 450-600/km^2 for total land, and on the order of 600-800/km^2 for cropped land. For each man, woman, and child resident in Sansai there is about 0.69 *rai* (0.11 hectare) of crop land in use. However, much of the land is used for two, or even three annual crops, so that the effective area per person is somewhat greater. If we consider "total

Table 6 *Population Density in Saraphi District, Sansai Subdistrict, and Sansai**

Political Division	Total Area	Cropped Area	Population (1969)	Population Density (Total Area, km^2)	Population Density (Cropped Area, km^2)
Saraphi district	127.3	102.3	60,000	471	587
Sansai subdistrict	11.9	9.5	5,315	446	559
Sansai	1.34	1.07	905	675	846

* The data presented are based on several estimates. The estimate of Sansai Subdistrict's population in 1969 is based on a count of houses in all villages other than Sansai; this figure was then multiplied by 4.87 (average household size in Sansai). "Cropped area" figures for Saraphi district and Sansai subdistrict were estimated. Estimates for the district and subdistrict were based on the assumption that the ratio cropped land:total land is the same as in Sansai.

land used" Sansai households during all three agricultural seasons, the figure per person is 1.21 *rai* (0.20 hectare).

In addition to a large human population, village lands also support a large number of domestic animals. At the time the village was surveyed in October, 1969, village households owned 77 water buffaloes, 5 bullocks, 503 hogs, 2290 chickens, and 352 ducks. During the previous year, villagers had sold 17 buffaloes, in most cases to other residents of the village. They also sold 693 hogs, 1057 chickens, and 131 ducks. In nearly all cases, hogs and poultry were sold to dealers from outside the village.

Socioeconomic Variation

About three months into the research period, a questionnaire was administered in each of the village households. It was administered in cooperation with the Chulalongkorn University Social Science Research Institute (CUSSRI) and the National Economic Development Board of Thailand. The same questions were asked in approximately 3000 households in the northern, central, and northeastern regions of Thailand. It took roughly ninety minutes to administer the *Socio-Economic Profile Schedule* in each household. Information was requested on household composition, occupations, employment, landholding, farming, capital equipment, financial arrangements, and related topics. The data provide rich insights into socioeconomic differences which exist within the village.

For purposes of analysis, the village was divided into subgroups through application of a socioeconomic rating system. The rating system was applied to the data of the CUSSRI *Socio-Economic Profile Schedule,* and is similar to that used by Oscar Lewis in the Mexican village of Tepoztlan in the 1940's (1963:173). In general, each point on the scale corresponds to 500 to 600 *baht* of annual income. A few items were included (e.g. schooling and size of labor force) for which it is difficult to assign a cash value. Yet it is felt that they are useful in differentiating socioeconomic groups. Points were assigned as follows:

	Points
For each *rai* of house land	1
For each longan tree	0.1
For each *rai* of field land owned, and used during a given season for rice, garlic, or watermelons	2
For each *rai* of field land owned, and used during a given season for peanuts or soybeans	1
For each *rai* of field land rented out, and used for peanuts or soybeans	1
For each *rai* of field land rented out, and used for rice, watermelons, or garlic	1
For each *rai* of field land rented in, and used for rice, watermelons, or garlic	1
For each *rai* of field land rented in, and used for peanuts or soybeans	0.5
For each member of the household labor force	1
For each year of schooling of a household member beyond Grade Four	0.25
For each 100 *baht* of monthly income income from a salaried job	2
For each 100 *baht* of monthly gross income from market selling or a small business	1
For each 100 *baht* of monthly gross income from a shop	0.25
For maintaining a savings account	1
For borrowing money from a credit cooperative or financial institution	2
For a buffalo	2
For an adult pig, two small pigs, twenty ducks, or fifty chickens	1
For a rice mill, large truck, or tractor	10

	Points
For a small truck	5
For an ox cart	1
For a motorcycle	1

Socioeconomic ratings for village households ranged from a minimum of zero to a maximum 179 points. The village was divided into three socioeconomic groups, of equal size, by assigning the 62 households with highest scores to Group I, the next 62 households to Group II, and the remaining 62 households to Group III. Equivalence between these groups and social classes should not be assumed. They are significant in this case because they facilitate analysis of quantitative data and demonstrate relationships between relative wealth and agricultural strategies.

The validity of dividing Sansai into socioeconomic groups according to the standards outlined previously can be tested through observation of household spending patterns. Data on household expenditures were gathered in a separate survey. A total of thirty-one households were surveyed--ten each from Groups I and II, and 11 from Group III. Table 7 summarizes data gathered in the survey.

The survey was made only once. The respondent for each household was asked to remember purchases for the entire previous year. Therefore, it is essential to remember that the data used here are estimates, subject to personality quirks of individual respondents. Nevertheless, the aggregated data demonstrate consistent differences between the Groups in the direction expected. Group I households spend more on nearly all items. Each of the expenditure items included in Table 7 will be considered in turn.

Housekeeping necessities. This category includes such items as pots and pans, flatware, dishes, glasses, furniture, bedding, lampfuel, batteries, mats, and washing powder. The survey shows precise ordering, whether one considers per capita values, household averages, or median household values. On an individual basis, members of Group I households spent an average fifty-one percent more on these items than households in Group III.

Medicine and medical treatment. The variety of medicines included in Table 7 is potentially wide. The category encompasses native herb medicines, Chinese medicines, various patent medicines, and western medicines. Medical treatment is provided by physicians at government and private hospitals, by injection doctors, and by a variety of traditional practitioners. If we consider per capita figures, household averages, and median household values, the rank ordering is not perfect in this case. In fact, there are indications that, among the households surveyed, the poorest spend most on medicines and medical treatment. A number of explanations are possible. The poor may, in fact, be ill more often, due to poor nutrition and other factors. Some families undoubtly *become* poor because they have one or more sickly members.

It may be that the poor adopt a unique life style which involves few trips to Chiang Mai or Lamphun cities (where most physicians are located) and great dependence on traditional medical practitioners and injection doctors. Medical treatment of the latter type turns out to be expensive in individual cases. In cases of chronic disease, the injection doctor arrives to give an injection every day or two for weeks or months at a time. There are cases at Sansai where patients of one particular injection doctor have lost their land as a result of an extended illness.

Merit making. In Sansai, merit making activities center around feeding the monks, giving them gifts, and the construction of religious buildings.

Table 7 *Annual Household Expenditures for Sansai**

Expenditure	Group I			Group II			Group III		
	Per Capita Average	House-hold Average	Median House-hold Value	Per Capita Average	House-hold Average	Median House-hold Value	Per Capita Average	House-hold Average	Median House-hold Value
1. Housekeeping necessities	94	550	620	78	380	410	62	280	220
2. Medicine and medical treatment	44	260	200	44	210	140	110	490	210
3. Merit making	68	400	340	39	190	140	73	320	190
4. Clothing and personal grooming	250	1,480	1,380	74	360	120	76	340	260
5. Entertainment	200	1,160	990	120	580	630	81	360	360
6. Food [2]	830	4,880	4,310	640	3,090	3,200	590	2,620	2,800
7. Total basic necessities	1,530	9,040	7,870	980	4,710	4,580	910	4,070	3,990
8. Total expenditures (including unusual expenses and capital expenditures)	3,440	20,400	17,000	1,450	6,940	6,080	1,290	5,770	6,280

* Items 1-6 cannot be summed to receive either the total under 7 or 8. Several expenditure categories--transportation, education, housing, farming expenses, and business expenses--are not included here in detail. To some extent they are included under item 7. *All* expenditures are included under item 8.

For purposes of this survey, donations for school construction, and gifts made at weddings and funerals were included in the same category. Once again, the data do not conform to the Group ordering expected. The appropriate explanation is that one household in Group III gave an unusually large amount (1,125) and that three households in Group II gave unusually small amounts (less than 100). In a larger sample, we can expect the usual Group ordering to prevail.

Clothing and personal grooming. Items included under personal grooming are: watches and jewelry, cosmetics, combs, brushes, soaps, toothpaste, haircuts, hairdressing, and related items. Group I households spend substantially more. The figures for Group II and III households are virtually the same. Household figures for Group II show particularly wide variation. Two households spend only 400 and 50 *baht*, respectively, while two other households spent 947 and 1,147 *baht*. These latter figures raise the household *average*, while the *median* value is quite low.

Entertainment. Expenses included under this heading are: purchase (and repair) of radios and similar appliances, newspapers and magazines, gambling losses, toys, tickets for movies, traditional plays, and temple festivals, liquor, and tobacco products. For most village households, the last item is the major expense under "entertainment." Some ready-made cigarettes are purchased, but most adult villagers smoke the traditional *bulii khijoo*. Banana leaf wrappers, bulk tobacco and *khijoo* (an aromatic ingredient made from tamarind bark) are purchased every few days. In some households, this single item represents more than ten percent of annual "basic" expenditures.

Surprising insight into village life is gained through a comparison of spending patterns for tobacco products and merit making. The data show that twenty-seven households spent more on tobacco products in 1969, only four spent more on merit making. Among villagers who consider themselves devout Buddhists, and for whom merit making is the most obvious path to improved future lives, 87.1 percent of households surveyed spent more on tobacco products. Expenditures on "entertainment" follow the usual Group ordering.

Food. The dietary patterns of village households are, in broad outline, similar. Rice provides the majority of the food energy, and of a number of other nutrients. Substantial quantities of vegetables are combined with small amounts of fish and meat products. Nevertheless, the data indicate that Group I households are able to spend more on food. On the average, Group I households spend forty-one percent more on food for each member than Group III households.

Total basic necessities. Included under this category are annual expenses that all households can expect. Items excluded are: business expenses, loan payments, and purchase of such items as tractors, hand tractors, trucks, land, houses, and other buildings, which are acquired occasionally, or once in a lifetime. Farming expenses are also excluded, although most families have them. They fill the needs of household members only indirectly. Once again, the data are in the expected Group order. Group I households spend sixty-eight percent more on ordinary consumer goods and services than Group III households.

Total expenditures. This category includes all items on which Sansai households spend money, including major capital expenditures and land. It is in these figures that we see greatest differences between Groups. Wealthy households have a substantially higher living standard than poor ones, but it is in major expenditures--on house, land, farming, and businesses--where differences are greatest. For instance, four households in Group I spent 3,200 - 13,000 *baht* on building new houses or improving old ones. No household in Group II or III spent more than 800 *baht* for the same purpose. For Group I households, an average 55.5 percent of total spending was devoted to "unusual expenses," business expenses, and capital expenditures. For Groups II and III the figures respectively were 32.4 and 28.7 percent. The wealthier households are able to devote more of their earnings to those items which will assure them greater earnings and security in the future.

Two other indicators, not included in Table 7, also separate wealthy households from the poor. Under present conditions, only well-to-do Sansai households are able to provide their children with education beyond the four years offered at the village school. Doing so requires that children be sent to stay either in Chiang Mai or Lamphun, and the expense involved is substantial. As a result, only one of the households included in the survey--a Group I household--had a child attending school outside the village. The cost of supporting that child in school for a year was 2,700 *baht*. No household in Group II or III spent more than 90 *baht* on a child attending the local school.

An indicator of living standard often applied in underdeveloped countries is the ratio of household food expenditures to total "ordinary" expenditures. In Sansai, food represents fifty-four percent of expenditures on basic necessities in Group I, and sixty-six and sixty-four percent in Groups II and III, respectively. This is yet another indicator that Group I households are accustomed to a higher living standard.

Villagers are inclined to stress their essential uniformity in contrast to the wealth and power of government officials and rich city dwellers. The opposite tack has been adopted in this study. Data have been marshalled which indicate that *substantial* socioeconomic variation characterizes life styles at Sansai. For purposes of discussion, Sansai's 186 households have been divided into three Socioeconomic Groups. This simple division has limited relationship to a complex social reality, but insights are provided. Differences between Groups are numerous, and statistically significant. They vary in economic strategies and life style.

Financially, households in Group I vary from moderately well-to-do to rich. All trucks, hand tractors, motorcycles, and rice mills are owned by members of this Group. Only these households are able to provide their children with advanced schooling outside the village. Sansai's few permanent, full-time landlords belong to this Group. They control an inordinate proportion of both house and field land. Only individual members of Group I have been able to purchase land at Sansai in recent years. Several households own large, profitable longan orchards within the village.

In addition to land, these households possess an inordinate share of locally significant occupational skills. The most successful middlemen, seamstresses, shopowners, and market women belong to Group I including both local injection doctors. It is from these households that leaders of local government, the irrigation societies, and the temple congregation ordinarily are chosen. Households in this Group are more likely than those in the other Groups to maintain savings accounts, to make small loans to friends, relatives, and neighbors, and to receive low interest loans (twelve percent per annum) from government or private sources. Members of Group I own and sell more livestock than members of other Groups.

The bulk of Group II households are firmly involved in agriculture. Most operate small, owned plots, or larger, rented plots. Members of many Group II households work at agricultural labor, but this constitutes a secondary source of income. Children in these households stand to inherit very small, if any, amounts of land. There are few signs of upward mobility. Children cannot attend school beyond Grade Four, and households are unable to purchase new land. Members of the Group work at occupations that require traditional, rather than "modern," white-collar skills. They work as carpenters, sawyers, basket-makers, mat-weavers, etc. The skills involved may be substantial, but financial rewards are consistently low. Nevertheless, these households are relatively well off by comparison with those in Group III.

In an agricultural area where ownership of land is the most obvious manifestation of wealth, most households in Group III own *no* land. Self-employed, farming households work rented plots on a half-share basis. Many households receive the bulk of their income from agricultural employment. This is true although agricultural laborers are ordinarily hired on a day-to-day basis. The wages they receive are four to seven *baht* per day.

Several households in Group III consist of elderly widows, living alone, and subsisting on donations provided by relatives and neighbors. Children in this group do not remain in school past Grade Four. No member of this Group has been chosen a *wat* congregation, irrigation society, village, or Subdistrict leader in recent memory. Such leaders are almost always chosen from Group I, and only occasionally from Group II.

Group III households do not maintain savings accounts, or receive low interest loans from government or private sources. Their only potential source of productive capital is a cooperative landlord who may be willing to pay full price for fertilizer, seed, and insecticide at the beginning of the season. He receives in exchange a half-share of a garlic or melon crop, and repayment of half his capital expenditure after the crop is sold. Local money lenders are used by members of this Group who extract interest payments of 3-5 percent per month, and require title to land as security.

The Annual Cycle

In an agrarian society, knowledge of the amount and type of labor expended month-by-month provides important insights into the annual cycle. While the annual cycle may have crucial sociocultural facets, this approach strongly emphasizes the importance of economic factors. Data on "labor expenditure and the annual cycle" presented in this section provide an important basis for analysis of strategies in crop choice at Sansai.

Data appropriate to analysis of the annual cycle at Sansai were gathered in ten households. From September, 1969 to August, 1970, one member of each household was asked to fill out, each day, a form which solicits information on the work schedule of each member of the household labor force. Completed forms were collected, and new forms distributed, once or twice each week. (See Table 8).

The ten households used in this "sample" were chosen before adequate socioeconomic data were gathered. In consequence, the sample does not include equal numbers of households from the three Socioeconomic Groups (See Table 9). Group I is too heavily- and Group III too lightly-represented. Nevertheless, all three Groups are represented, and sample households fall close to village norms so far as a number of economic activities are concerned.

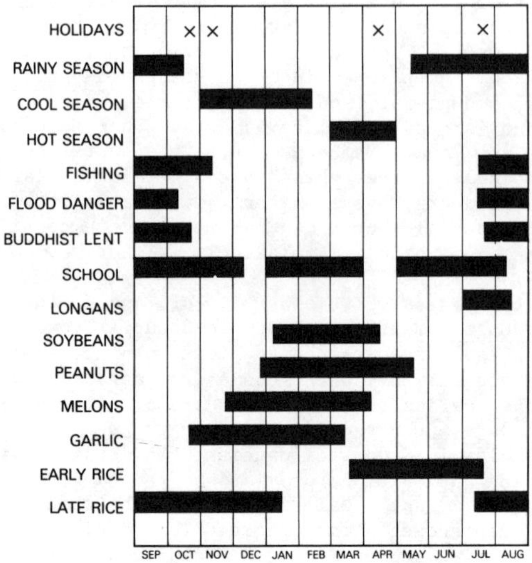

Table 8 *The Annual Cycle at Sansai*

For example, during the year spanning April, 1968 - March, 1969, 51.1 percent of *all* village households raised the early rice crop, 64.5 percent raised the late rice crop, and 73.1 percent raised the garden crop. For the ten sample households, the equivalent figures for the year spanning September, 1969 - August, 1970, are 50, 60, and 90 percent, respectively. Forty percent of the sample households own water buffalo, compared with 34.9 of all village households. One hundred percent of the sample households own pigs and poultry, compared with 85.5 percent of all village households that own pigs, and 82.8 percent that own poultry. Where crop production and care of livestock are concerned, the sample households are likely to reflect village patterns of labor expenditure with fair accuracy.

The sample contains data on one full-time school teacher, one full-time research assistant, and one individual who worked full-time at teaching, and devoted weekends and vacations to assisting the writer in his research. These three individuals represent unusually heavy labor expenditures on "employment." Similarly, the sample may be skewed in the area of labor expenditure on "self-employment." In the *Socio-Economic Profile Schedule*, ı embers of 35.5 percent of all village households admitted to working at self-employed tasks. For the sample group the figure is 80 percent.

The sample group can be considered fairly representative of the entire village where concern is with labor expenditures on cropping, care of livestock, fishing and chores. It is not fairly representative of the entire village so far as labor inputs into employment and self-employment are concerned.

According to Table 9 average "labor years" vary from slightly over 1,000 hours per capita to nearly 3,000 hours. This substantial range is explained by variation in labor force composition, by variation in age and

Table 9 *Data on Ten Sansai Households that Recorded Daily Schedules From September, 1969 to August, 1970*

House-hold	Socio-economic Group	Number of Fulltime Workers[3]	Total Hours Worked	Hours Per Worker	Raise Early Rice?	Raise Late Rice?	Raise Garden Crop?	Own Water Buffalo?	Employed More than One Month?	Self-Employed More than One Month?
A	I	3 1/4	7,360	2,260	-	-	X	X	X	X
B	I	3 1/3	6,405	1,930	-	-	X	-	X	-
C	I	2	5,780	2,930	-	-	X	-	X	X
D	III	3 1/12	4,455	1,450	-	-	-	-	X	X
E	I	7	11,520	1,650	X	X	X	X	-	X
F	II	3	6,510	2,170	X	X	X	-	-	-
G	II	3 2/3	4,015	1,090	X	X	X	X	-	X
H	I	3 3/4	6,665	1,780	-	X	X	-	-	X
I	II	4 2/3	7,350	1,570	X	X	X	X	-	X
J	III	2	5,635	2,820	X	X	X	-	-	X

health of individual workers, by variation in amount of land owned and worked, and by possession or lack of occupational skills outside the agricultural sector.

Chart 4 presents data on total labor expenditures for the ten households in graphic form. The image of labor expenditure presented for a northern Thai "rice" village is striking. Of total labor expended by ten households, only one-tenth is devoted to tilling, sowing, transplanting, irrigating, harvesting, threshing, transporting, storing, and marketing the rice crops! An ordinary farming household in this "rice" village is likely to devote more time to cultivation of garden crops, to care of livestock, or to cooking and related activities, than to rice cultivation.

The planting of three annual crops in some plots at Sansai is not unique in the Chiang Mai Plain. Kingshill reports similar practice at Ku Daeng, seven kilometers away. At least some Ku Daeng fields are planted with early and late rice and a third crop of soybeans, tobacco, peanuts, or vegetables (Kingshill 1965:28-41). The cropping schedule at Sansai probably is broadly representative of a number of densely settled, irrigated areas of the Chiang Mai Plain. Although subject to numerous minor variations in individual cases, the annual cropping cycle at Sansai takes the following general form.

April: Those farmers who raise the early rice crop begin to prepare their fields for transplanting. The iron-tipped plow and the wooden-toothed harrow ordinarily are drawn by a single water buffalo. After plowing and harrowing the field is sometimes "puddled"; the buffalo draws a heavy board over the muddy field. When the field is finally ready to receive the transplanted seedlings, it is covered by a virtually level layer of smooth mud, 10-15 centimeters thick. Late in the month some farmers begin transplanting seedlings from a nursery into the prepared fields. The seedlings are planted by hand in the smooth mud. Those farmers who planted peanuts earlier in the year are now harvesting.

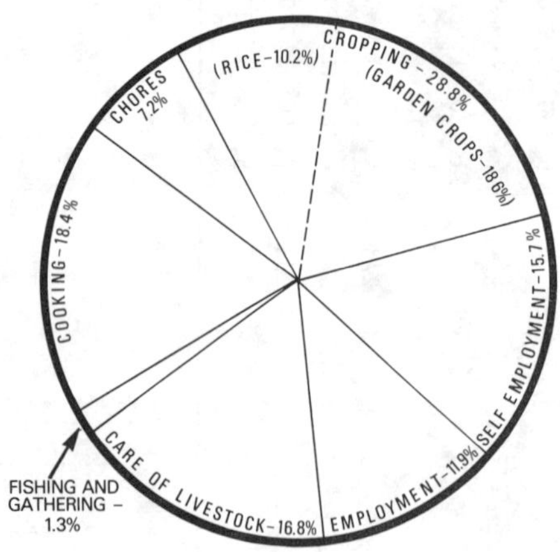

Cropping (Rice 6,700 hours) (Garden crops 12,170 hours)	18,870 hours
Self-employment	10,300 hours
Employment	7,800 hours
Care of livestock	11,040 hours
Fishing/gathering	880 hours
Cooking, etc.	12,040 hours
Household chores	4,740 hours
Total	65,600 hours

Chart 4 *Total Labor Hours Expended by Ten Sansai Households During a Twelve Month Period, 1969-1970*

May: Transplanting of the early rice crop continues. When this task is completed, a relatively inactive phase of the agricultural cycle is reached. It may be necessary to do occasional weeding of the recently transplanted rice. Otherwise, farmers must concern themselves only with maintaining an appropriate water level in their flooded paddy fields.

June: The tasks of weeding and irrigation are continued through the month. At the end of the month a few farmers begin to harvest early rice.

July: Harvest of the early rice crop continues throughout the month. The entire process--cutting, threshing, winnowing, and transport to storage--is carried out with only the assistance of simple hand tools. Immediately after harvest, many fields are plowed and harrowed in preparation for transplanting of seedlings from nurseries prepared and sowed earlier in the month. Late in the month the annual longan harvest begins, producing fruit of high value. Most owners sell the fruit, while it is still on the tree, to a local middleman.

August: The longan harvest continues into the middle of the month. Transplanting of the late rice crop begins early in the month. This is a period of considerable tension for local farmers. In most years there is a threat of flooding at this time. If flood waters do rise, and do so soon after the seedlings are transplanted, most or all of the plants will drown. When this happens, as it often does, farmers must wait until the flood waters recede. They then devote much time to preparing and sowing a new nursery or to locating seedlings for purchase. If the floods arrive, farmers can expect to spend considerable time guarding and repairing the levee which protects their fields from devastation by the swollen Ping River.

October: For those farmers cultivating late rice, October is also an inactive month, punctuated only by occasional weeding and irrigation. For a number of farmers who do not raise this crop, preparation of fields for garlic planting is begun. Because the pace of field preparation can be leisurely, plowing and harrowing usually are done with buffalo-drawn implements at this time.

November: Garlic is planted throughout the month. However, early in the month it is planted only in those fields that were fallow during the late rice cropping season. In mid-month, harvest of late rice begins. For many Sansai farmers this initiates the busiest period of the annual cycle which extends into February or March. Rice harvest is followed immediately by preparation of the fields for annual garden crops. Many farmers find it convenient to hire tractors to plow at this time. This tactic is relatively expensive, but it saves a few days at a time when both household and paid labor supplies are severely restricted.

December: The late rice harvest, plowing, harrowing, and preparation of the land for particular garden crops continues through the month. Garlic is planted until mid-month; after this most farmers plant watermelons and soybeans.

January: Planting of melons and soybeans continues and farmers also begin to plant peanuts. After planting, watermelons must be hand-watered at two or three day intervals. Garlic and peanuts are irrigated at fifteen day intervals. Occasional weeding may also be necessary. Those farmers forced by flooding to transplant late rice a second time begin to harvest at the end of the month.

February: The "belated" rice harvest continues until mid-month. Weeding, watering, and irrigation of garden crops continues. Late in the month the garlic harvest is begun. The farmers who harvest earliest usually expect to sell their product fresh to middlemen representing garlic pickling firms who come to the village.

March: The garlic harvest continues. By mid-month most harvested garlic is being dried and stored for sale later in the year. Watermelons are harvested; some residents are absent from the village for several days, retailing melons at a number of towns and cities in northern Thailand. Soybeans are harvested, steamed, and sold as snacks at several nearby markets. After harvest of garlic, melons, and soybeans, some fields are immediately prepared as nurseries for the early rice crop. Sowing takes place near the end of the month.

The labor supply at Sansai is subject to substantial pressure during part of the year. The two major garden crops in 1970 were garlic and watermelons. Both require heavy inputs of labor, and households raising two or three *rai* of one or both find themselves working very hard from late November until March. For example, a member of Household A devoted more than 1,000 hours to cultivation of melons between November, 1969 and March, 1970. Two members of Household C devoted more than 1,500 hours to garlic and melon crops during the same period. During these months the couple also devoted more than 400 hours to employed and self-employed tasks, and additional hours to live stock care, household chores, and cooking and marketing.

Table 10 offers a different view of labor expenditure at Sansai. Items in Line A demonstrate that the pattern of agricultural labor expenditure is *strongly* cyclical. There are distinct peaks and valleys. In the busiest month the surveyed households worked fifteen times as many hours as in the slowest month. Slack periods are September-October and May-June. These periods follow planting of the two rice crops, when there is little to do in the fields beyond occasional weeding and water control. Between the September-October slack period and the May-June slack period, there are six months of intense activity. During this time late rice is harvested, melons, garlic, soybeans, and peanuts are cultivated and sold, and early rice is sowed in the nurseries and transplanted. Between the May-June and September-October periods there is a shorter period of intense activity devoted to harvest of early rice, sale and harvest of longans, and sowing and transplanting of late rice.

Items in lines B and C (Table 10) provide insights into labor expenditure on self-employment and employment at Sansai, and the relationship between these and labor expended on agriculture. The "valleys"--months in which relatively few hours are spent on self-employed tasks--correspond to the months of maximum agricultural activity. For the ten surveyed households, the three busiest months are December, January, and March. It is precisely these three months that are most "slack" for self-employed tasks. For many households there is also a lull in self-employment in July-August, when one crop must be harvested, and another transplanted, simultaneously.

Employment is not a major source of income for most households at Sansai. Only five or six households have members who are employed full-time, year round. Since two of these households are included in the sample, it is not truly representative of village households in the area of employment. Because the bulk of employed labor conforms to the requirements of the Thai government and two anthropological researchers, the annual cycle of employed labor does not bear any obvious relationship to the cropping cycle.

Table 10 *Labor Hours Devoted to Various Activities by Ten Sansai Households During a Twelve Month Period*

ACTIVITY	SEP	OCT	NOV	DEC	JAN	FEB	MAR	APR	MAY	JUNE	JULY	AUG
A. Agriculture	455	240	1215	3645	2865	2105	2530	1160	400	635	1955	1665
(Rice)	(400)	(100)	(590)	(735)	(0)	(0)	(110)	(595)	(195)	(560)	(1900)	(1495)
(Garden Crops)	(55)	(140)	(625)	(2910)	(2865)	(2105)	(2420)	(565)	(205)	(45)	(55)	(170)
B. Self-employment	840	760	890	650	620	845	575	910	945	1105	1270	890
(Commerce)	(260)	(250)	(180)	(180)	(180)	(370)	(260)	(340)	(255)	(420)	(585)	(520)
(Other)	(580)	(510)	(710)	(470)	(440)	(475)	(315)	(570)	(690)	(685)	(685)	(370)
C. Employment	680	745	640	700	825	645	575	320	590	585	775	720
(Teaching/Research)	(615)	(600)	(530)	(480)	(535)	(510)	(420)	(285)	(550)	(530)	(595)	(505)
(Other)	(65)	(145)	(110)	(220)	(290)	(135)	(155)	(35)	(40)	(55)	(180)	(215)
D. Care of livestock	765	785	915	640	855	820	815	1030	1220	1085	1090	1015
(Cattle/Buffalo	(245)	(245)	(245)	(90)	(255)	(290)	(365)	(495)	(745)	(545)	(615)	(615)
(Pigs/Poultry)	(520)	(540)	(670)	(550)	(600)	(530)	(450)	(535)	(475)	(540)	(475)	(400)
E. Fishing and food gathering	445	150	55	2	7	0	0	5	19	38	12	130
F. Cooking, marketing, and related tasks	990	1085	955	965	1120	835	810	1010	1060	1065	1040	1100
G. Household chores	500	760	480	305	260	275	260	365	590	325	315	305
(Ordinary chores)	(210)	(295)	(265)	(305)	(260)	(275)	(260)	(325)	(320)	(325)	(315)	(305)
(House construction-Household F)	(290)	(465)	(215)	(0)	(0)	(0)	(0)	(40)	(270)	(0)	(0)	(0)

Line D (Table 10) is concerned with labor inputs into the care of livestock. It is apparent that the busiest month for cropping--December--is the least busy for care of livestock. Labor inputs into care of water buffalo are necessarily heavy. Sansai's dense population allows for no areas of free grazing. The animals are either fed rice straw and cut grass, or allowed to graze in the school yard, cremation grounds, or fallow fields under direct supervision. Variation in labor expenditure on care of these animals is explainable in terms of variation in supplies of rice straw and grass for cutting, and in terms of labor requirements of field crops.

Labor inputs into pig and poultry care vary relatively little, and according to subtle patterns, during the year. Chickens need little care. It requires a few minutes a day to scatter a few handfuls of milled rice. Ducks require substantial care only when they are young. For a period of weeks, ducklings are driven to water to feed on aquatic plants. Later, they are given unmilled rice and are able to reach the water on their own. Hogs are given weeds, banana stalks, and rice bran. The first two items are gathered, while bran is purchased at one of the local rice mills. Most households prefer to feed bran when the demand for agricultural labor is great, and weeds and banana stalks when excess household labor is available. In addition, the supply of weeds reaches a peak in the latter portion of the monsoon rains, and a low point in April.

Line E (Table 10) provides data on fishing and food gathering at Sansai. Villagers actually devote little time to food gathering. Two elderly women spent a total of seventeen hours digging for a species of burrowing cricket that is found in the village. Ordinarily, these crickets, along with other insects, snails, and freshwater crabs, are gathered by children.

Individuals old enough to belong to the labor force in the ten sample households devoted more than 850 hours to fishing during the survey year. The great bulk of this activity was restricted to the months of August, September, and October. This is a period when flood waters around the village are high, fishing is good, and people have substantial amounts of leisure time. No fishing is done by working members of the sample households during February and March, and little during December, January, and April. Some fishing is done in November, May, June, and July. The availability of fresh fish in the daily markets reflects this pattern of labor inputs. Very few individuals sell fresh fish during February, March, and April. During the period from August to November there are many sellers.

Line F (Table 10) is concerned with labor inputs into cooking, marketing, dishwashing, woodchopping and other activities directly related to food preparation.[4] Labor inputs are substantial (more time is spent on food preparation than on rice production) and there is little variation from month to month.

Line G (Table 10) provides data on labor inputs into household chores. These chores, relatively simple and not particularly time-consuming, include: sweeping and cleaning within the house, sweeping the bare ground immediately surrounding the house, washing clothes, filling water pots, and occasional construction of fences and other minor projects. Labor inputs into ordinary chores vary little from month to month.

Household F built a new, wooden house during the year. Workers in the household included the household head and his wife, both in their early 30's, and a son in his early teens. During the slack period of September-November, the three devoted more than 950 hours to house construction. The incomplete house was left standing during the garden crop season. After garden crops were harvested, an additional investment of 310 hours in late April and May was sufficient to complete the dwelling.

Chapter Four

THE VILLAGE: SOCIOCULTURAL DIMENSIONS OF AGRICULTURAL DECISION-MAKING

Crucial decisions at Sansai and elsewhere are effected within constraints imposed by the culture, social system, and environment. Chapter Three provides insights into constraints of the last type. Major demographic, edaphic, and climatological features of the Sansai eco-system are specified. Sansai's population, densely settled, manifests a high dependency ratio. Local soils are fresh water alluvium. They are regularly enriched by silts washed down from the mountains.

The annual climatic cycle of Sansai includes distinct monsoon, cool, and hot seasons. Rainfall is moderate, and highly variable at the beginning and end of the monsoon rains. Major features of local agriculture are: patterns of double- and triple-cropping; a labor cycle which manifests modest "rest" periods in May-June and August-September; and heavy work schedules for most households during the rest of the year. One rice crop is raised largely for local consumption. A second rice crop and several garden crops are grown for sale.

This Chapter provides a general description of the sociocultural milieu within which agricultural decisions are made. In Shimkin's terminology, the problem is one of specification of "operational postulates." His discussion of "playing the games of survival" suggests several appropriate descriptive and analytical foci (1966:45-47). In describing agricultural decision-making at Sansai it is useful to consider: sources and techniques for information gathering; payoff functions; the value of pure versus mixed strategies; procedures for evaluation and self-correction; patterns of organization; and the role of "cheating."

These topics, and others, will be considered in this Chapter under the following headings: Sansai agricultural terminology; decision-making skills; information gathering; evaluation of available strategies; social organization, culture, and decision-making; and adaptive strategies.

Sansai Agricultural Terminology

Terms used by local residents provide important insights into agricultural choice. In some cases they delimit and distinguish entire decision frameworks. In others they delimit and distinguish particular strategies. In still others they provide indices suitable for measuring or predicting payoffs. Of significance to the present discussion are distinctions between the following paired terms: *naa* and *suan*; *dau* and *pii*; *chaw* and *baeng kyng*; *kaa kaan* and *kaa khwaaj*; *tang* and *thaang*; and *haj myang* and *haj phasii*.

Naa. This noun denotes a set of agricultural techniques applied to a piece of land. The Malay term *sawah* is equivalenc (Geertz, 1963). A piece of land is called *naa* if it has been divided into earth-walled sections and if each section has been more or less carefully leveled. *Naa* must also have irrigation facilities sufficient to deliver standing water during at least part of the cropping season. Only wet rice is raised in *naa*. No land at Sansai is used only for *naa*.

Suan. Virtually all other crops are raised in *suan*. *Suan* are orchards (e.g. longan orchards), dry-cropped fields (planted in tobacco or soybeans), and partially flooded or non-flooded, irrigated fields planted in melons, peanuts, or garlic. Most fields at Sansai are utilized as *naa* during one or two annual cropping seasons, and as *suan* during the remaining season. *Suan* plots are sometimes subjected to mixed use. Garlic and rice seedlings are often cultivated between rows of immature longan trees. A few cabbages are raised in garlic *suan*. Most *suan* crops at Sansai are cultivated between November and May. Nevertheless, the fruit of the longan--harvested in July and August--are a *suan* crop.

Dau. This modifier denotes relative "earliness," e.g. longan varieties harvested in July are called *dau*, while those picked later are called *pii*. Similarly, *khaaw* (rice) *dau* is harvested in June-July and *khaaw pii* (late rice) in November-December. But the "earliness" indicated is always relative to some specific category. One variety of *khaaw pii* is *khaaw kaew dau*, indicating that it matures relatively early for a "late" variety.

Pii. This modifier denotes relative "lateness " When applied to the rice crop at Sansai *pii* carries several additional connotations. The early and late rice crops are planted in distinct varieties, raised for different purposes. These varieties, until about fifty years ago, were taxed differently. *Naa dau* are planted in nonglutinous varieties which enter national and international marketing networks. *Naa pii* are planted in glutinous varieties which are the mainstay of the local diet. In the late nineteenth century, when Sansai was ruled directly by *Caaw* Mahawong of the Chiang Mai line, the two rice crops were taxed differently. Local commoners were required to cultivate a plot in the "royal fields" and also permitted to cultivate private lands. During *naa dau* the *Caaw* received one-half of the produce of the royal field and one-third of the produce of private lands. During *naa pii*, the entire production of the royal field was received by the *Caaw*, and private fields were not taxed [1]

Chaw. At Sansai it is possible to "rent" (*chaw*) land, draft animals, or agricultural equipment. The rental fee for the first and third items is invariably paid in cash. Owners of rented cattle or buffalo ordinarily accept payment in paddy. At Sansai only *suan* lands are rented. Thus tenancy arrangements involving cash payments are contracted only in November-December. During the research period rents were nearly always set at 100 *baht* per *rai*. The landlord who accepts a rent surrenders control over crop choice. However, he is freed from the responsibility to provide seed, fertilizer, insecticide, hired labor, a draft aminal, or a hired tractor for cultivation.

Baeng kyng. A literal translation of these words is "divide in half." No general term denoting "sharecropping" or "sharecropper" was recorded in

Sansai. Sharecropping traditionally has operated, and continues to operate, on a half-share basis. Tenants raising rice always are subjected to this arrangement, and it enjoys equal popularity with cash rental during the *suan* season. Owner and tenant each receive one-half of the produce at harvest. Flexibility is introduced into the arrangement largely through negotiation regarding *kaa kaan* and *kaa khwaaj*.

Kaa kaan. This noun denotes an obligation associated with a piece of land. The owner or operator must participate in the maintenance of the irrigation system which serves the land. This obligation to participate in manual labor, or to hire a substitute, falls automatically on the owner-operator. In verbal contracts between tenant and landlord, it becomes a major bargaining point. One of the pair accepts *kaa kaan*, while the other accepts *kaa khwaaj*. Major landowners who accept *kaa kaan* ordinarily hire a replacement, in some cases the tenant.

Kaa khwaaj. This phrase means literally the "obligation of the water buffalo." In negotiations between landlord and tenant, the individual who does not accept *kaa kaan* usually accepts *kaa khwaaj*. Traditionally, this implied provision of the draft animal with which the tenant plowed and harrowed the sharecropped plot. Recently, major landlords accepting this obligation have sometimes hired a tractor and driver to plow at a cost of 40-80 *baht* per *rai*.

Tang. This noun denotes a standardized volume used to measure crop yields at Sansai. The *tang* is a factory-produced wooden bucket of 20 liter capacity. When filled "level full" with paddy, its capacity is 10.2 kilograms (22.4 pounds). It is nearly always used when tenant and landlord divide a harvest.

Thaang. This noun denotes the ubiquitous carrying basket that does not provide a standard volume measure. The capacity of baskets used locally varies narrowly around 1.5 *tang*, or 15 kilograms of paddy. The *thaang* is sometimes used to measure rice yield by owner-operators who are not concerned with extreme accuracy.

Haj myang. The official standard for measurement of land area in Thailand is the *rai* (0.16 hectare). Northern Thai has a related term, *haj*, which also refers to a unit of land. But the area is not standardized. The term refers to an earth-walled plot of a size "typical" of such plots locally. At Sansai this approximates 0.25 *rai*. The informality and lack of precision in the term is expressed clearly in the term *haj myang* which translated literally means "native *rai*."

Haj phasii. Where collection of precise data is required, use of the modifier *phasii* is crucial. In nominal form it denotes "taxes." This phrase can be translated literally as "tax *rai*." It connotes "standardized land measure of the Central Government," and it is equivalent to *rai* in Central Thai.

Ngaam. This modifier denotes "prettiness." In agricultural contexts it is used to describe a field crop consisting of large, strong, and healthy plants.

Bau ngaam. In agricultural contexts this phrase is applied to stands of plants that are small, weak, and unhealthy. This term, and its negations, are frequently used by Sansai residents in accounting for selection or rejection of a particular crop or variety.

Utilizing these, and related, terms, Sansai residents define the domain of agricultural decision-making. *Naa dau*, *naa pii*, and *suan* constitute three distinct cropping seasons, and three distinct, but interrelated, decision frameworks. *Chaw* and *baeng kyng* define alternate strategies in acquiring land for garden crops. *Tang* and *haj phasii* in combination constitute a precise, modern index of crop yield. *Thaang* and *haj myang* in combination provide a cruder, more traditional, but serviceable index for measurement of payoff function. *Ngaam* and *bau ngaam* provide an index for measurement of plant quality.

Decision-making Skills

Since the 1930's, Sansai has had a four year public school. Education beyond four grades requires attending a school outside the village. The mode for the number of years of education completed by Sansai adults is four. As a result, the literacy rate in Central Thai is high. Most villagers can read a newspaper, write a letter, and do simple arithmetic. Elderly men who are unschooled in the Central Thai language often are literate in the Northern Thai script.

Sansai residents, as a group, possess skills potentially applicable to record-keeping and agricultural decision-making. What they lack is training in formal bookkeeping. Villagers occasionally keep financial records, e.g. contributions made to a temple building fund or for a cremation or wedding. But *no* local resident, including the major landowners, applies bookkeeping techniques to cropping activities. If rational decisions--regarding timing, crop and varietal choice, use of fertilizers and insecticides--are made, substantial feats of memory and calculation are implied.

If we consider only the skills involved in evaluation of competing crops or varieties, complexity is substantial. The farmer is called upon to remember his yields and expenditures, and those of his neighbors as well. He must remember these figures for several years at a time, and for a number of crops and varieties. From this mass of data he extracts or abstracts the crucial payoff functions upon which judgments are based. He must reconcile the experiences of "people in general" with his particular experiences. If he is truly rational, he must consider the probability and probable effect of exceptionally good or bad weather conditions.

However, in laying bare the complexity of skills required, it is inappropriate to assume that they are beyond the reach of Sansai residents. The other side of the coin requires examination. Sansai residents display an intense, continuing interest in the cost of things. It is virtually impossible to purchase a new item of equipment or clothing without being asked several times "How much did it cost?" or "What arrangements were made for payment?"

Data on labor costs and agricultural yields seem to be of equal interest. The outsider eventually gains the impression that input/output data are being processed constantly, although on an informal (without paper or pencil) basis. One illustration is the cooperative labor groups. They ordinarily operate over the length of a cropping season. Representatives of several households cooperate in a variety of tasks, but particularly in the tasks of transplanting, weeding, and harvest. When the appropriate time arrives, all households with participating members expect to receive labor directly equivalent (in man-days) to that donated. Such groups typically include five or ten households, and involve the donation of dozens or hundreds of days of labor. Different household members may serve at different times. Yet the system operates with sufficient accuracy and equity so that group membership sometimes remains stable for several years at a time.

One local resident does record payments and receipts in his work as a longan middleman. His educational level (five years) is slightly above the village mode. Recorded data are of real value to him, because single year contracts are arranged several weeks before harvest, and in a few cases, complex, multi-year contracts are negotiated. He does not keep books for the other commodities--garlic and peanuts--in which he deals. With them, all payments and receipts related to a particular transaction are handled within several days. The general shops at Sansai operate, sometimes quite successfully, without the benefit of general bookkeeping. Two or three shopkeepers do record credit purchases made by a small number of favored households, but beyond this no records are kept.

The Central Government does provide some training in bookkeeping procedures. Farmers receiving loans from the Bank for Agriculture and Agricultural Cooperatives for raising hogs are encouraged to record expenditures on food, medicine, etc. But during the period of fieldwork, application of these techniques did not spread. Farmers required to keep records did so for the activities supported by bank loans. However, the techniques were not applied to other agricultural or business activities in the same household, and the practice had not spread to other households.

A series of informal interviews were carried out during March, 1970, with the purpose of exposing major features of decision-making for the annual garden crop. Although farmers were somewhat diffident in explaining their reasoning, their efforts to achieve improved profits clearly were evident:

Mr. Chan, House #54

This year, (1969/70) he planted ½ *rai* of watermelons. Although he does not own the land he was allowed its free use by the owner. The field is part of the land that he sharecrops for early and late rice each year. He can plant any garden crop he chooses. He decided to plant melons because he heard his neighbors talking about their high profits last year. Last year he planted 1¼ *rai* of garlic, and 1 *rai* of peanuts. The year before that he planted all garlic. He changes from crop to crop because the soil gets poor if he replants the same crop every year. Next year he will plant both garlic and melons to "hedge his bets," or perhaps he will "depend on his fate." His wife assists him in raising the crop. He hired a tractor to plow the field for 40 *baht*. He spent 230 *baht* on "red" and "white" fertilizer. His major problem is that the soil gets poorer if he plants the same crop every year.

Mr. Bud, House #40/2

This year he planted 1½ *rai* of melons and 3/4 *rai* of garlic. Both were planted on land belonging to his father who lives in the same house. He planted the same amount of garlic the last two years. Melons are new this year. He planted them because he heard his neighbors saying that they were the best crop to raise locally. His production of melons this year was "average" or "ordinary." He hasn't decided vet whether he will plant garlic or melons next year. He needs to know the size of the local melon crop first (i.e. how many farmers are planning to plant the crop). His nephew and he together work the land. He hired a tractor to plow for 100 *baht* and spent 230 *baht* on "white" fertilizer. The major "problem" with planting melons is that it is difficult and takes a great deal of care. He planted two crops this year, to "hedge his bets." As it turned out, garlic did better than melons (i.e. received a better price) because there were many melons produced this year.

Mr. Kaew, House #80/1

This year he planted ½ *rai* of garlic and 1 *rai* of melons. The melons were planted on owned land, while the garlic land was rented for 50 *baht*. He decided to plant these two crops because his neighbors were doing the same. Last year he planted 3/4 *rai* of peanuts and ½ *rai* of garlic on his own land. This year he changed to melons because last year there was too little water and the peanuts did poorly. He plants melons in the higher plots because he is afraid that excess moisture will harm them. Next year he will plant only garlic. He does not want to plant melons because he must spend a lot on fertilizer. Garlic is more attractive because the price was not too low this year, and he did not lose money on it. This year he planted melons because he was following his neighbors. His wife assists him. He spent 240 *baht* on "white" and "red" fertilizer. His major problems are plant diseases and the worms that destroyed some melons this year. He spent 123 *baht* on three brands of insecticide. In his opinion, the best month to plant garden crops is November. There are no days of the week when it is particularly "auspicious" to plant.

While local residents generally explain their choice procedures briefly, and in simplistic terms, it seem clear that numerous strategies are evaluated, and that substantial effort is devoted to the evaluation process. Mixed strategies are applied, e.g. garlic-peanuts or garlic-melons. Farmers also explain with considerable clarity that this simultaneously lessens potential profit *and* potential loss. Mr. Pan explains this concept clearly in the following interview:

Mr. Pan, House #78

This year he planted one *rai* each of melons and of garlic on rented land. He planted *both* crops because he figures that if the price of one was down, the other probably would be up.

Mr. Chan (#54) is able to verbalize his approval of the concept of crop rotation. In the following interview, Mr. Suu outlines an interesting approach to rotating crops:

Mr. Suu, House #22

This year Mr. Suu planted both garlic and melons on rented land. He plans to plant the same crops again next year, but it will be necessary to find new land to rent. As a result of planting the same crops on the land now rented during two consecutive garden seasons, production has already begun to fall off. This year's melon crop is clearly inferior to that of last year. Some plants have died, and the watermelons produced are small.

Because land is relatively easily rented during the garden crop season, Mr. Suu is able to achieve the effect of crop rotation by renting land used previously for some garden crop other than melons or garlic.

Mr. Bud (#40/2) is aware of the heavy labor inputs required by the melon crop, and also that his yield was no better than "average." Mr. Kaew objects to melons because of the heavy expenditure required for fertilizer and insecticide. All three men display a willingness to adopt new crops and techniques. All three give evidence of considering available evidence collected from their neighbors, and from the annual "experiments" carried out on their own lands. Although they are diffident in explaining their decision processes, it is obvious that they work hard at collecting data, testing strategies, and making the best available choice.

Information Gathering

Sansai residents appear to collect agricultural data from two major sources: personal experiments and neighbors. Interviews with Chan, Bud, Kaew, and dozens of other local farmers attest to this statement. Local farmers are willing innovators, but information sources which result in, or influence, change, are largely personal and local. A strategy is adopted or rejected on the basis of the farmer's *personal* experience, and/or the experiences of his fellow villagers. In the fields, temple, or market, on the path or road, during evening strolls, valuable information is exchanged on proper techniques, costs, yields, and prices received. There are few barriers to free exchange of data. No castes, corporate kin groups, or permanent factions inhibit its flow. There are, to be sure, ties of kinship, friendship, and neighborhood. Yet none of these bonds seem strong enough significantly to stem or redirect the flow of information. If a village resident enjoys great success with a new rice variety, it is probable that all Sansai farmers will hear about it a few weeks after harvest.

In farming, Sansai residents deal directly with *local* soils and climate, and with *local* irrigation and transport facilities and labor supply. The most significant test of crop, variety, technique, or equipment is under local conditions. Information flows within the village link up the results of many local experiments. An interview with Mr. Phaun demonstrates a willingness to experiment:

Mr. Phaun, House #17

This year he planted ½ *rai* of melons, and ¼ *rai* of garlic. Next year, he will plant only garlic. This year the price of melons is too low. He feels that soybeans would not do well on his land. He knows this because his father planted soybeans on the adjoining land this year, and they were *bau ngaam*. After harvesting the garlic and melons, he will plant *phakkaad* (a variety of cabbage) on the land. He does not feel that there will be enough water for early rice this year and wants to experiment with this new crop.

While oral communication within the village and the immediate locality probably is the most significant source of agricultural data, residents have access to additional, outside information sources. A sizable minority of local households own transistor radios, and have access to daily agricultural news. Five stations are received in the village; two of them regularly broadcast *kam myang* (Northern Thai) programs. The 12:30 p.m. broadcast of agricultural news presented daily in Northern Thai was particularly popular. Programming consisted of information on wholesale commodity prices, weather forecasts, warnings of flood danger, and occasional information on new techniques and equipment. These data are interspersed with reports on temple festivals, local politics, and related subjects.

Some Sansai residents who are members of "credit clubs" have access to another source of data. The clubs are organized by the Bank for Agriculture and Agricultural Cooperatives. They consist of 15-20 members who act as co-guarantors for agricultural loans. Once accepted, each member applies for a separate loan. The bank requires a security deposit, usually in the form of a title to land. If a club member defaults, other members are responsible for repayment. Bank officials attempt to avoid default through provision of relevant information. Loans are given to buy or improve land, a water pump, hand tractor, or draft animal. The money received may be spent on digging a well or fish pond, or to buying hogs, fertilizer, or high quality seed. In the context of club meetings, members are provided with substantial information on modern agricultural techniques. The following brief record of a club meeting provides an example:

> (August 16, 1969.)
>
> At the organizational meeting of the "Credit Club" at Sansai, Mr. Phaun, a bank official, spent several minutes explaining interest rates, repayment schedules, and required security deposits with potential members. Then he gave some guidance as to how the loans might be used. He mentioned some experiments with a new variety of glutinous rice in progress in Sansai District (not the same as Sansai Sub-district). He listed some new fertilizers that were on the market. Several "old" club members asked for additional information. Then Mr. Tomsii, who had accompanied Mr. Phaun, gave a talk on modern techniques of hog raising. He had been trained at the Agricultural Training School at Mae Coo. He spoke at some length about the new, manufactured hog feeds that recently had become available.

Mr. Phaun is a dedicated and hard-working official. He was to be seen in the village two or three times each month, meeting with members of the credit club individually and in groups. Occasionally, a new rice variety, fertilizer, or pig feed was recommended. During the research period, 12-14 Sansai residents belonged to credit clubs, and were recipients of such advice. There is little evidence that information received in this matter was widely or quickly diffused during the research period.

Recommendations offered by a government official, no matter how much he is trusted and liked, lack concreteness and comparability in the local situation. A rice variety with a high yield in central Thailand may do poorly under local conditions. For example, its growing season may be too long to fit successfully into a pattern of triple-cropping. A promising new fertilizer may be unavailable locally, too costly, or untested on locally raised rice varieties. Recommendations made by government officials must often be regarded as little more than "potential strategies, untested under local conditions." Initially, a certain amount of inertia has to be overcome. If the recommended variety, technique, or product *is* tested locally, and performs better than existing varieties, techniques, or products, information will spread rapidly. But this is the result of concreteness and comparability lent to the recommendation by local experiments.

The three Ministries of Agriculture, Rice, and Irrigation maintain experiment stations at which new agricultural strategies are developed and tested. But officials of these Ministries who work at the District level provide an insignificant source of agricultural information. They visit in the village less often than once a year and seldom confront the farmers face to face. They do appear occasionally at monthly meetings for sub district and village headmen held at the District office. But any information received from these officials by Sansai farmers is filtered down through two local officials with little personal interest in agriculture.

Villagers also tend to accept advice from government officials with a grain of salt. In 1969, officials of the Rice Experiment Station in San Pratong District recommended moderate fertilizer applications with a new rice variety. They predicted yields of 80-100 buckets per *rai*. Sansai residents ignored advice regarding the use of fertilizer and received a creditable sixty buckets per *rai*.

Another incident demonstrates the caution and humor with which Sansai residents react to government programs. District officials came to the village during the first week of research to vaccinate draft animals against hemorrhagic septicemia. The program was free, voluntary, and had been carried out successfully during two previous years. Several villagers arrived with their cattle and buffalo at 9:30 a.m., and quickly set up two wooden fences to be used as a restraining device. They then waited more than two hours for the government veterinarian to arrive. Agriculturally, the season was a busy one. Most of the men had pressing field work to do. While waiting, the men displayed an attitude of good-humored caution, their consistent attitude toward Central Government programs and officials. They laughed and joked that "The Government has called us here today to give our bodies a much needed rest."

Perhaps the most effective source of extra-village agricultural information is the fertilizer companies, particularly ESSO. ESSO officials appear in Sansai and nearby villages at regular intervals, to give public or semi-public presentations. They work closely with officials of the Bangkok Bank who operate credit clubs similar in format to those of the Bank of Agriculture and the Agricultural Cooperatives. The primary goal of ESSO public presentations is sales of fertilizer and insecticide. Yet company representatives provide information on crops, varieties, and techniques which combine successfully with their products. Frequently free samples accompany public presentations.

Evaluation of Available Strategies

Sansai residents are in a position to evaluate only those agricultural strategies which are locally available. No doubt, numerous other cropping strategies would be appropriate to the Sansai area. But only if appropriate seeds, fertilizer, equipment, and techniques are locally available. In fact, the range of available strategies is rather narrowly defined. During the research period, Sansai farmers chose among two early rice varieties, three late rice varieties, and four garden crops. Each range of choices is to some extent defined by conventional wisdom. Residents recognize that each crop or variety has certain general characteristics under local conditions. (Specific qualities which govern final choice are considered in Chapters Five, Six, and Seven.) Table 11 defines some general characteristics of garden crops raised at Sansai:

Table 11 *Characteristics of Garden Crops Raised at Sansai*

Crop	Labor Costs	Capital Costs	Maximum Return	Profit Variability/ Risk of Loss
Garlic	High	High	High	High
Watermelons	High	High	High	High
Peanuts	High	Low	Medium	Low
Soybeans	Low	Low	Low	Low

No crop at Sansai can be judged "best." Each has characteristics which make it attractive to some households, and unattractive to others. The resources of the household (in land, labor, capital, ready cash, skills, and social influence) affect their evaluation of these crops. Garlic is a high cost, high risk, but potentially high profit crop. It requires heavy labor inputs that are relatively concentrated. Often it is necessary to hire labor crews during planting and harvest. This is impossible without ready cash.

Melons are also a high cost, high risk, and potentially high profit crop. However, labor requirements are more evenly distributed over the season for garlic than for melons. The major problem with melons lies in marketing. No middleman network has developed so the producer must transport his crop to a spot suitable for retail sales. Ready access to a rented truck, whether through ready cash or influence with a truck owner, is crucial. Households lacking this access may be inclined to evaluate the melon strategy with great caution.

Garlic is a crop with a long history at Sansai. It has been planted for more than sixty years, and, until recently, it was cultivated without fertilizer. A few farmers still choose to raise this crop without fertilizer. This lowers the cost and risk involved, but also severely limits maximum expectable profits. Peanuts are a crop potentially attractive to households with a plentiful labor supply and limited cash resources. Soybeans represent small inputs of labor and cash. A few Sansai farmers find it expedient to plant them "speculatively." Soybeans are planted in full knowledge that if the supply of irrigation water in March-April appears to be adequate, the soybean vines will be plowed under for the early rice crop. Soybeans make no real sense on rented land. Average earnings and the usual rental price (100 *baht* per *rai*) nearly coincide.

An interview with Mr. Waaj demonstrates the thought processes which some local residents apply to agricultural choice:

> *Mr. Waaj, House #17*
>
> This year he planted 2 *rai* of garlic (on owned land), 1¼ *rai* of melons, and 1¼ *rai* of soybeans on land rented from Mrs. Peng who lives in a village in Lamphun Province. He pays her a cash rent which reflects the level of production. If production is poor, he may pay as little as 10 *baht* per *rai*. He reasons that *if* he had planted only melons, he would have encountered problems with marketing the crop, as has now become apparent. By planting a variety of crops, he lessens the associated risk. The soybeans planted this year are *bau ngaam* (not "pretty") and he will not plant this crop next year. He plans to continue planting some garlic because he does not want to "waste" land by planting some less profitable crop. He points out a piece of land (about 1¼ *rai*) and estimates that if planted in peanuts, it will produce about 180 *baht*. However, the land will produce about 400 *baht* planted in melons. If garlic received a price as high as it did several years ago (as much as four *baht* per kilogram), then garlic might earn considerably more than 700 *baht*.

Mixed strategies are common at Sansai. Mixing crops with diverse characteristics often has the effect of improving efficiency of household resource utilization. For example, households ordinarily lack sufficient labor resources to raise more than two or three *rai* of melons or garlic. It is common for households with six or eight *rai* of land, and limited labor supply, to put approximately two *rai* into garlic or melons and the remaining land into soybeans. This achieves a higher degree of "fit" between land and labor resources than any alternative which leaves land lying fallow. In addition, the area planted in garlic or melons benefits from chemical fertilization, while the area planted in soybeans profits from the nitrogen-fixing characteristics of that crop.

Although both are high cost, high risk crops, melons and garlic frequently are combined. There are at least two advantages to this mixed strategy. Labor requirements are redistributed. There is also a modest degree of "hedging" against possible price fluctuations. Combining melons or garlic with peanuts is a reasonable choice for households with ample labor, but limited funds for capital expenditures.

Mixed strategies are also common during the late rice cropping season. The two varieties most often combined have complementary strengths and weaknesses. The first variety planted on lower plots has high resistance to drowning (plants can survive up to ten days submerged) but is relatively slow to mature (early to middle December). The second variety planted on higher plots has little resistance to drowning (plants often die after four days submerged) but matures early (middle to late November). The second variety is harvested early, allowing early planting of garlic and melons (and higher prices). The first variety achieves a relatively high yield on low, flood-prone plots.

One last example reflects the significance of *availability* of agricultural strategies. Rice fields through large areas of the Chiang Mai plain gradually are being converted to longan orchards. But this represents a strategy *not* available to most farmers. The associated risks and access costs are too high.[2] Labor and cash expenditures involved are small, and the long range profit potential is large. The crux of the problem lies with the risk involved in taking land almost completely out of production for an unknown period of years. Longan trees seldom bear before the fifth year, and often produce little for eight or ten years. Later, after the first eight or ten years, longan orchards have a high potential for profit. But only well-to-do landowners and businessmen can afford to accept low or no profits in the beginning. For other farmers, the "access price" is too high so conversion of rice fields to longan orchards is not an available strategy. The majority of Sansai households can afford to plant only a few longan trees in their house compounds.

Sansai residents do not all encounter the same set of agricultural strategies. Neither do they evaluate available strategies from the same viewpoint.

Social Organization, Culture, and Decision-making

Sansai residents belong to groups, enter into relationships, and subscribe to beliefs and values which, potentially at least, constrain agricultural choice. A few of these constraints were made clear in interviews and informal discussions. Potential constraints which frequently were discussed include: obligations associated with the landlord-tenant relationship; obligations between individuals with adjoining fields; and magico-religious beliefs governing the timing of agricultural operations. Finally other features of Sansai social organization and culture of potential, but undemonstrated, significance will be briefly discussed.

The Landlord-Tenant Relationship

The nature of landlord-tenant interaction varies with the agricultural season, tenancy arrangement, and individual personalities. The following excerpts from interviews provide some insight into the variety associated with these arrangements:

> *Mr. Khwaaj, House #43*
> This year he rented 3/4 *rai* from Mr. Taa. He planted ½ *rai* of garlic and ¼ *rai* of peanuts on a half-share basis. Mr. Taa, as the landowner, chose the crops to be planted. Mr. Khwaaj is unable to make this choice since he owns no land.
>
> *Mr. Waan, House #88*
> Mr. Waan rents from Mr. Duang, the wealthiest landowner in Sansai. This year, by choice of Mr. Duang, he planted garlic. Expenses and profits are to be shared equally by landlord and tenant.

Mr. Thaun, House #123

Mr. Thaun planted 2½ *rai* of late rice on rented land. As a result of flooding in late August, all the plants drowned. The owner of the land, Mr. Maa, gave him 40 *baht* to buy seedlings sufficient to replant the area. (Mr. Maa also did the same for Mr. Thaun's brothers.)

Mr. Pan, House #78

Mr. Pan rents land owned by Sansai's Buddhist temple. The floods of late August (1969) destroyed about one-half of his crop. He has already purchased replacement seedlings in nearby villages. The cost so far has been about 30 *baht* but more seedlings are needed. The temple committee is reimbursing him for his expenditures.

Mr. Pan, House #102

Three *rai* rented from Mr. Duang were submerged for 15 days and all the plants drowned. After it had become apparent that the loss would be complete, the owner provided new seed. He also gave Mr. Pan permission to sow a crop of seedlings between the trees in his longan orchard. The seeds were sown 13 days ago. In 12 more days they will be pulled and transplanted. Mr. Pan also plants early rice on the land owned by Mr. Duang. He has done so only since 1968 when Mr. Duang first provided an irrigation pump. Previously, the land received too little water for successful cultivation during this season.

Important features of the landlord-tenant relationship are displayed or alluded to in the preceding excerpts. Mr. Khwaaj and Mr. Waan are both sharecroppers and work with major local landowners. Frequently, in such cases, the landlord selects the rice variety or garden crop. The few absentee landlords who own land at Sansai ordinarily allow sharecroppers to make their own selections. Two or three local residents who rent from Sansai's temple are able to make their own selections. Individuals who rent land on a cash basis to raise garden crops are, to some extent, independent decision-makers. However, owners can control crop choice by refusing to rent to individuals who make undesirable choices.

Although sharecroppers often forfeit substantial decision-making powers, they make substantial gains in security. The landlord is no less committed than the tenant to achieving a successful crop. For example, Mr. Thaun and Mr. Pan are assisted in repairing flood damaged fields. Mr. Pan is provided additional assistance in the form of an irrigation pump.

Obligations Between Farming "Neighbors"

Under some circumstances, operation of adjoining or neighboring fields imposes a degree of conformity on Sansai farmers. This is most obvious during the garden crop season when crops have variable water requirements. The following examples are illustrative:

Mr. Dii, House #87

This year he planted four *rai* of watermelons on rented land. Last year he planted ½ *rai* of garlic and one *rai* of peanuts on portions of the same land. He changed to melons this year more as a matter of necessity than of choice. Since owners of adjoining and nearby fields had chosen this crop, he had little choice but to follow suit. Had he chosen garlic or peanuts, it would have been impossible to irrigate the crops after planting. Next year, if the operators of adjoining fields change crops, he will do the same. For example, if neighboring farmers choose to raise garlic, he will be able to raise garlic, peanuts, or soybeans. If he *insisted* on planting garlic (and irrigating) while his neighbors plant melons, there would be legal trouble.

Mr. Thaun, House #6

He rents land from Mrs. Pan at 100 *baht* per *rai*. This year he planted two *rai* of melons and ½ *rai* of garlic. He decided to plant melons because they brought a good price last year. He planted ½ *rai* of garlic because the area planted is too low and damp for melons. He also had seed left from last year. He is forced to concentrate on melons because most of the land receives inadequate water at this time of year for peanuts, soybeans, or garlic. In addition, his watermelons are nearly surrounded by other fields in the same crop so it would be impossible to irrigate the land. If he did so, and melons in adjoining fields were damaged by seepage, he would be forced to reimburse the affected farmers for the damage.

Watermelons, a new crop in Sansai, are sensitive to excess moisture. Because of the high potential profits involved, both the operator and operators of adjoining plots must be alert to the possibility of damage. Individuals who operate land which is isolated from irrigation canals or ditches have little choice but to raise melons when their neighbors choose to do so. The following passage provides a slightly different example of imposed "neighborliness":

Mr. Duangcan, House #65

Mr. Duangcan produces early rice, late rice, and garden crops on a field owned by his mother. Beginning in 1968, an irrigation pump was used to provide water for early rice. He is co-owner of the pump with Mr. Bunthaa and Mr. Caj who operate nearby fields. The necessary well was dug in Mr. Duangcan's field. The operators of three other nearby fields were persuaded to use the services of the pump at a cost of 10 buckets per *rai* for the season. Thus, the operators of six adjoining fields are forced to work (in tandem.)

Several irrigation pumps are operated at Sansai to insure adequate water for the early rice crop. The adjoining plots of six to ten operators are ordinarily served. To a considerable extent, these operators must conform to a single schedule. Water usually reaches the lowest parts of the lowest field only after flowing through all the higher fields. This process is highly disruptive unless all operators choose to adjust water levels at the same time. By utilizing the services of an irrigation pump, Sansai residents agree to be governed by group decisions regarding the timing of agricultural operations.

Another aspect of intensified "neighborliness" results from the growing use of hired tractors for plowing garden plots. A farmer can seldom hire a tractor on his own. Tractor operators insist on plowing several *rai* at one time. Most farmers operate too little land to entice tractor operators to drive several kilometers for a single job. Tractor operators, enjoying a sellers' market during the period when melons and garlic are planted, normally expect to deal with the operators of several adjoining fields. There is also the problem of access. Unless the farmer is fortunate to operate a field directly adjoining the road, he must seek the permission of his neighbors to permit the hired tractor to pass through *their* fields. Obviously, permission will not be given if the field is planted. Many Sansai farmers forfeit individual control over cropping schedule when they choose to hire a tractor.

Magico-astrological Beliefs

All Sansai residents are aware of magico-astrological systems which purport to assure the quality and quantity of agricultural production. Some individuals have memorized a list of auspicious days for planting, transplanting, and other agricultural operations. Others consult fortune tellers or astrological charts. While some residents claim great faith in these techniques, others express strong skepticism. A variety of approaches are presented in the following accounts:

Mr. Kaung, House #51/1

Auspicious days for planting are Friday and Saturday. He knows this to be the case because he often has planted according to this formula.

Mr. Unryan, House #53

Auspicious days for planting are Thursday, Friday, and Monday.

Mr. Suk, House #18

This year he planted on rented land two *rai* of garlic, and two *rai* of soybeans. The soybeans did not do well this year (*bau ngaam*). Next year, he will plant garlic again on two *rai*. But he has not decided yet what to do with the other land. He will wait until the Northern Thai New Year celebration in April. At this time he will buy a printed chart from an astrologer at Nong Tong or Chiang Mai. An illustration on the chart, when read correctly, predicts the price of agricultural commodities and the amount of precipitation for the coming year. Prices are not predicted in terms of specific crops. Instead, there are obscure references to "things white," "things red," and "things green."

Mr. Thaun, House #6

In general, Monday and Tuesday are good days to plant. Saturday is a bad day. Monday is particularly auspicious for an individual born on Sunday. Tuesday is auspicious for an individual born on Monday. If the farmer is unable to plant on a day which is auspicious for him, he should find someone to plant who was born on the correct day.

Mr. Syyn, House #56/1

This year Mr. Syyn planted three *rai* of garlic, ½ *rai* of peanuts, and 100 melon plants (about 1/8 *rai*) on his own land. He rented his remaining land to a farmer at 100 *baht* per *rai*. He is not interested in auspicious days for planting. He does suggest that it is wise to plant in November to receive early season, high prices.

Mr. Maa, House #31

Mr. Maa had planted garlic for several years. As a young man, he experimented with a variety of crops, including soybeans and peanuts. He estimates the following income from ¼ *rai* of the following crops: peanuts, 150-250 *baht* and garlic, 400-600 *baht*. Mr. Maa concluded by saying that success in raising garden crops sometimes is a matter of fate and rainfall.

Although knowledge of magico-religious beliefs is widespread in Sansai, it is not clear that such beliefs are a significant constraint on rational decision-making in agriculture. Some residents are openly skeptcal. Others claim to be true believers, but manifest hardheaded practicality in their actual behavior. In addition, some of the schemes followed may be no less "rational" than other randomization devices.

Mr. Kaung, Mr. Unryan, and Mr. Thaun all subscribe to the notion that there are auspicious and inauspicious days for carrying out agricultural operations. Yet it is inappropriate to assume that these beliefs restrain farmers from doing everything practically possible to achieve a successful crop. While Mr. Kaung may transplant only on Friday and Saturday, he does so only after careful consideration of practical matters:

Mr. Kaung, House #51/1

Mr. Kaung planted five *rai* of late rice on land belonging to his mother. The August flood caused complete destruction to his plants. Mr. Kaung is faced with the task of replanting. Unlike many other Sansai farmers, he does not need to purchase seedlings. He will replant this field with plants taken from another, higher field belonging to the household. This three *rai* field was transplanted, at double density, about two months ago. The plants are about 45 centimeters tall. Members of the household will go through the field pulling up every second plant. Sufficient plants will be collected from the three *rai* field to replant five *rai*. This technique of double-density planting for insurance purposes is called *kaa cam* or *kaa bong*.

Mr. Syyn expresses his skepticism of these beliefs. Mr. Maa explains in magico-astrological terms ("fate") what is essentially a mathematical and scientific principle; commodity prices and climate at Sansai are highly unpredictable, although probablilites can be assigned to various possible outcomes. Mr. Suk's reliance on astrological charts probably is best interpreted as a rational technique for randomization in the face of highly unpredictable climatic and market conditions. At the beginning of a season, when positive (e.g. high commodity prices and adequate water supply) and negative (e.g. low commodity prices and too much or too little water) outcomes may seem equally likely, the charts help him to adopt *some* strategy. Where the probability exists that either outcome (positive or negative) strongly outweighs the other, one assumes that Mr. Suk does not refer to astrological charts.

Other Sociocultural Features of Potential Significance

Other features of Sansai social organization and culture have potential effect on agricultural choice. Among these features are: general obligations associated with community membership; household composition; membership in an irrigation society or a cooperative labor group; the relationship between borrower and lender; dietary preferences; and certain features of "outside" social organization.

Residence in a northern Thai village imposes a set of general social obligations on household members. Adult males sometimes are asked to accept community offices. Households are expected to provide workers for village construction and cleanup projects. They are expected to participate in weddings, funerals, temple and school festivals, and similar public events. Such obligations are potentially significant to the extent that they interfere with agricultural tasks. But their actual significance, in this sense, probably is small. Full-time farmers are seldom elected or appointed to public office at Sansai. During the research period, all such offices were held by non-farming landowners, shopkeepers, and middlemen. Work projects--road repairs or cleaning the cremation ground of underbrush--occasionally are organized at times which are agriculturally inopportune. But such projects seldom last longer than half a day. Any adult or teenage household member can fulfill the labor requirement.

While attendance at weddings, funerals, temple festivals, and similar events is important, when work in the fields is particularly pressing it takes precedence. For example, the holiday which initiates the Buddhist "Lent" (*waan khaaw phansaa*) is important at Sansai. In 1969 it occurred at the end of July, during a period when many farmers had to harvest early rice and transplant late rice in quick succession. Households who were particularly busy at this time spent the day in the fields, not in the temple.

For many adult male residents, obligations of citizenship and fellowship extend to participation in gambling activities. Except during "Lent," cockfights are held each week at a location close to Sansai. These events were not studied carefully, but it was apparent from occasional attendance that thousands of *baht* change hands at each fight. Rummy games, involving modest stakes, are common in the village, occurring in the village barber shop and homes. Sansai residents are also frequent participants in lotteries, both legal and illegal. Such activities affect agricultural choice either by providing unexpected cash to be used for capital expenditures or by unexpectedly removing a cash reserve. Although the subject was not studied carefully, gambling activities are responsible for a modest redistribution of productive resources at Sansai.

Ordinarily the Sansai nuclear family, the household, and agricultural production unit coincide. Exceptions are rather easily defined. In two or three cases, a pair of households cooperate in agricultural activities. In each case, one household is lineally descended from the other. During the garden season, households sometimes break into two or more farming units. Adult, unmarried sons, and newly married sons still living with the parents often choose to rent land and work it on their own.

This practice never occurs for either of the rice crops. The most obvious explanation for seasonal variation in household economic organization lies with the greater availability of rental land during the garden season.

Under other circumstances, the household can be considered the decision-making unit. If the household contains several adults, choice of crops, varieties, techniques, and related matters may involve complex interactions. In small households it is probably safe to attribute the role of major decision-maker to the household head, and to assume that wives and working children fill limited advisory roles.

All Sansai rice farmers belong to one or more irrigation societies. All fields within village boundaries are served by a single canal, and are associated with a single irrigation society. Residents of several other villages also belong to this society. Sansai residents who operate rice fields outside village boundaries sometimes belong to two other irrigation societies. Each society imposes contradictory constraints on its members. At some point, pure cooperation is in order, e.g. in maintenance of major canals and ditches or in petitioning Irrigation Department officials for early release of water. In other situations, small groups and individuals find themselves engaged in pure competition, e.g. when water is in short supply and needs are great, or when lowland farmers are simultaneously coping with too much and highland farmers with too little water. In recognition of the competitive nature of irrigation groups, irrigation headmen are conferred substantial powers of mediation and assessing fines.

Roughly one-half of Sansai's rice farming households participate in cooperative labor groups. Such groups ordinarily combine the labor resources of five or six households. The groups usually devote themselves to the tasks of transplanting, weeding, and harvest. Ideally, the groups provide household decision-makers with adequate labor when it is needed most. Since, however, such groups work for only one household at a time, membership may impose scheduling adjustments from time to time. A few days delay may be significant during harvest. The late rice crop is likely to lose moisture from plant and grain, resulting in broken stems and grains, and lower yield. When early rice is left standing too long in the field, the danger of flood damage must be considered.

Borrowing large sums of money ordinarily results in substantial constraints on agricultural choice. Small landowners ordinarily borrow from local moneylenders. A usual condition of such loans is temporary forfeiture of title and use of the agricultural land. Thus the range of strategies available to the borrower is substantially narrowed. He must either work with less (owned) land or seek out a rental or sharecropping arrangement. Frequently the arrangement is completed when the borrower purchases at a bargain price the land given as security. Several fields have changed hands in this manner.

Landowners eligible for low interest (one percent per month loans) from government and private sources must accept modest constraints with their loans. Members of borrowing clubs sponsored by the Bank for Agriculture and Agricultural Cooperatives and the Bangkok Bank are subject to regular supervision and advice. Undoubtedly farmers desiring to receive future loans will make an effort to please and impress bank officials. Occasionally, this may result in an alteration of customary techniques.

Perceptions by Sansai residents of the relative cash values of glutinous and nonglutinous rice varieties are affected by dietary preferences.

Since glutinous varieties are the local staple, and nonglutinous varieties are used only occasionally, the former should be considered "more valuable" on a unit basis. Nonglutinous varieties produced up to the value of household requirements probably should be assigned a cash value at retail levels. Nonglutinous varieties, which are largely exported from the village, should be assigned a cash value at wholesale levels.

Sansai residents demonstrate that this relationship holds true by their willingness to give outside traders twelve buckets of early, nonglutinous rice for ten buckets of late, glutinous rice. These trades are made each year, in spite of the fact that nonglutinous rice ordinarily receives a retail price in the village 10-25 percent higher than that of glutinous varieties. A trade at this ratio makes sense only if we realize that some Sansai households are unable to raise the late, glutinous crop. They are, in effect, trading a cash crop (at wholesale prices) for a staple food (at retail prices). In these terms the trade is "rational" as long as the current retail price of glutinous rice exceeds the current wholesale price of nonglutinous varieties by a factor greater than 1.2:1.

To a considerable extent, Sansai farmers are affected by extra-village social forces. Local land prices have reached such high levels that land purchase is no longer a strategy open to local farmers. Only school teachers, shop owners, and major, non-farming landowners can buy land. Two police officials who reside in Chiang Mai, but have kinsmen in the village, in recent years have purchased sizable plots.

Another external constraint of potential significance is the rice premium policy of the Thai government. The rice premium is a tax levied by the government on all rice exported from Thailand. Its effect is to maintain off-farm rice prices at roughly one-half of the international price (Silcock, 1970:19). It is sometimes argued that artificially low rice prices are a major constraint on use of fertilizer and other capital items (H.S.H. Prince Sithiporn, 1969:40).

Adaptive Strategies

Agricultural choice is a significant aspect of adaptation at Sansai. It has not yet been established that this aspect of adaptation is governed by rationality. However, the descriptive materials presented in this monograph suggest that local farmers are active, progressive decision-makers, within the constraints imposed by social organization and culture. They are neither passive nor conservative. Characteristics of adaptive strategies in Sansai agriculture will be summarized under the following headings: characteristics of players and habitat; patterns of organization; and operational postulates.[3]

Characteristics of Players and Habitats

Physiographic and edaphic variations within the Sansai ecosystem are minor. Within an area of 1.35 km^2 altitude varies no more than 1.5 meters, while alluvial soils prevail throughout the area. However, highly variable monsoon rains combine with a gravity-operated irrigation system to introduce substantial unreliability into the ecosystem. Farmers raising early rice risk crop damage or complete loss as a result of dessication. Farmers raising late rice must contend with unpredictable floods.

Local farm operators vary substantially in their power to make crucial decisions. Owner-operators ordinarily select rice varieties, garden crops, techniques, and timing with a substantial degree of independence. However, under specified circumstances when his neighbors raise watermelons, when he utilizes a pump for irrigation, when he belongs to a cooperative labor group--the range of available strategies may be narrowed. Rent-paying and sharecropping tenants ordinarily surrender part or all of their decision-making powers to the landlord.

Patterns of Organization

From some viewpoints, Sansai farmers form large, interlocking networks. This is especially true in the matter of information exchange. Information on annual informal "experiments" passes freely from farmer to farmer. Households combine their labor resources for greater efficiency. The services of irrigation pump and tractor frequently are utilized in tandem. But cooperative ventures frequently present anomalous facades. Local irrigation societies provide a simultaneous focus for cooperation and competition. Members must cooperate in maintenance tasks and in approaching Central Government authority. But intra-group relations sometimes are marked by fierce competition for limited water supplies. Competition for land also places individuals or households in direct competition. Landlords are able to choose among a substantial number of potential tenants. Disagreement regarding distribution of inherited land is a frequent cause of legal disputes between siblings.

Operational Postulates

Sansai decision-makers operate under the constant constraint of incomplete information. They must choose crops, varieties, and techniques *without* a full inventory of possible payoffs. Long established strategies may provide nearly complete information, but promise lower payoffs than newer, strategies for which less information exists. Sansai residents indicate clearly their understanding of the situation by frequent application of mixed strategies. For example, melons and garlic are raised together, as a hedge against unpredictable price swings. Farmers regard each year's planting pattern as an experiment to be altered without compunction if some other arrangement appears superior. "Cheating" is commonplace. Within recent decades several "rule changes"--new crops, new rice varieties, chemical fertilizers, tractors, and insecticide--have been accepted with enthusiasm.

In Chapters Three and Four the rules and constraints within which Sansai farmers operate have been described in a general, non-quantitative manner. In the process the nature of their agricultural goals (specifically their rational self-interest) has been presented anecdotally or merely assumed. In the remaining Chapters, analysis will consist of careful, quantitative examination of farming practice (Chapters Five and Six) and development of an explicit model of crop choice (Chapters Seven and Eight). The reader will find in these analyses convincing evidence that Sansai farmers are rational maximizers, motivated by material self-interest.

Chapter Five

STRATEGIES IN RICE CULTIVATION

Chapter Five begins with a description of the rice plant and those physiological requirements and characteristics which are most crucial within the Sansai environment. These data provide the background necessary for the interpretation of rice cropping strategies applied at Sansai. Maps, graphs, tables, and statistical analysis are used in describing these rice cropping strategies.

Two rice crops are raised at Sansai each year. They are called *naa dau* (early rice) and *naa pii* (late rice). Some fields, because of problems with water supply, are "too high" for early rice, and are used only for late rice. On the other hand, other fields are "too low" for late rice and are planted only with early rice. Many fields are of "medium" height, and used for both rice crops. (See Map 5.)

The seeds for the early crop are planted in nurseries in February and March. The seedlings are transplanted in March and April and harvested in June and July. The seedlings for the late crop, planted in nurseries between late May and October, are transplanted four to eight weeks later. The late crop is harvested between mid-November and early February. The early crop consists of nonglutinous varieties, the late crop of glutinous varieties.

Most of the data included in this Chapter are drawn from two surveys. One was carried out after the late rice crop of 1969, and the other after the early rice crop of 1970. Data were gathered on size of labor force, landholding status, size of holding, production expenses, variety used, labor inputs, and yields. The data were collected to provide a sound basis for delineating and evaluating strategies used by village rice farmers. Only village households who raise rice within village boundaries were surveyed.

To understand the problems which face the Sansai rice farmer, some knowledge of the rice plant is required.

The Rice Plant

This discussion of the rice plant focuses on: the distinction between glutinous and nonglutinous varieties; the distinction between photosensitive and non-photosensitive varieties; factors controlling plant growth and yield; factors regulating biological competition; and decisions "imposed" by the rice plant.

Map 5 *Estimated Altitude of Sansai Fields*

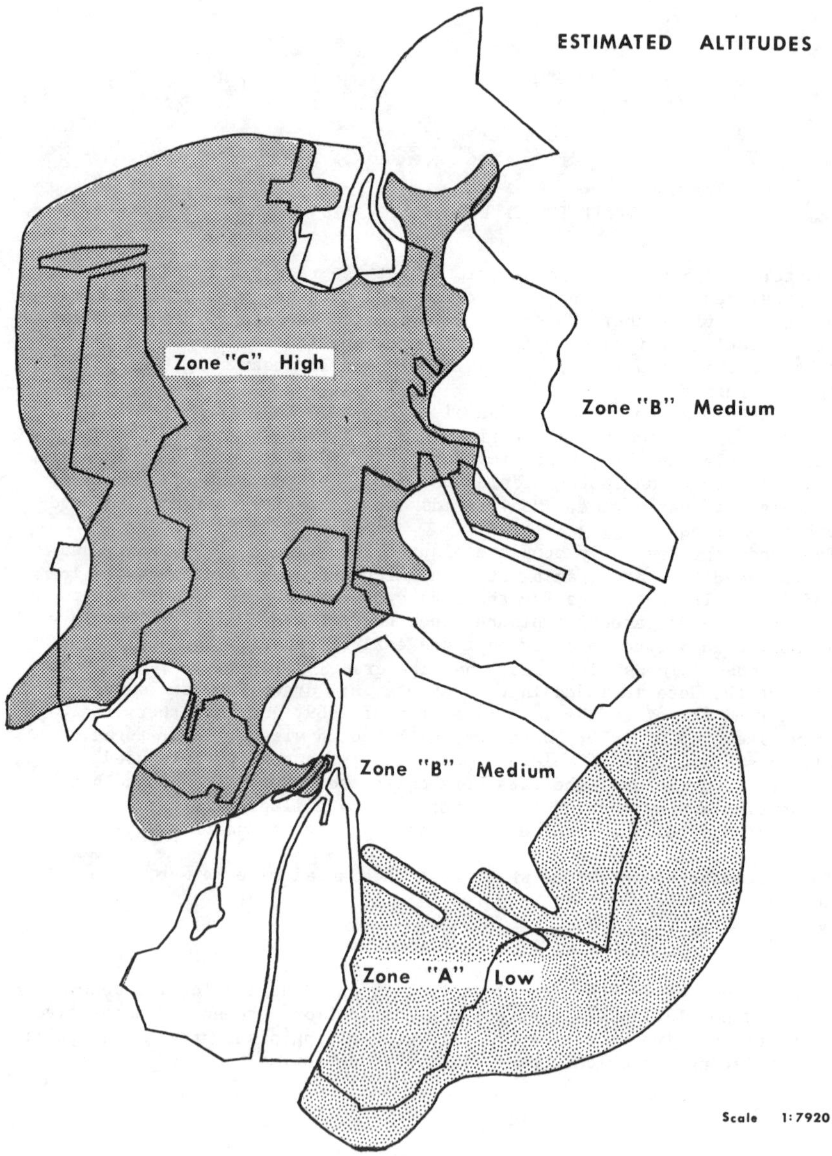

Glutinous and Nonglutinous Varieties

Crucial to understanding of agricultural decision-making at Sansai is the distinction between glutinous and nonglutinous rice varieties. Since the former are eaten daily in most Sansai households, and the latter are used only occasionally, the distinction is economically and agronomically significant, i.e. one between a "cash" and "subsistence" crop.

The two rice varieties can be distinguished informally by their cooking quality. Glutinous rice coheres into a sticky mass at temperatures as low as 150-160°F., while nonglutinous or ordinary rice remains relatively unaffected at these temperatures (Chandraratna 1964:162). Another difference is that ordinary rice stains blue in a potassium-iodide solution, while glutinous rice stains reddish brown (Chang and Bardenas 1965:11). Chemically, the difference between the two types is focused on the starch content of the endosperm, and seems to have two main dimensions.

The first difference is the presence or absence of dextrose. Glutinous rice contains four or five percent dextrose in the endosperm, while ordinary rice contains none. The second difference is the ratio of amylose to amylopectin. Glutinous rice contains almost no amylose, so that the amylopectin: amylose ratio is much higher than in ordinary rice. The glutinous character of the endosperm is thought to be determined by a single recessive allele (Chandraratna 1964:162).

Rice varieties of the two types do not differ significantly in content of major nutrients (United States Department of Agriculture 1963:52). However, some researchers are convinced that the daily consumption of glutinous rice in northern and northeastern Thailand contributes to the high rates of bladder stone disease that have been observed in those areas (Valyasevi, et al. 1967).

Photoperiod Sensitive and Non-photoperiod Sensitive Varieties

Another important distinction among rice varieties is that between photoperiod sensitive and non-photoperiod sensitive plants. Varieties raised for the early crop at Sansai are non-photoperiod sensitive, while varieties planted during the late rice season are photoperiod sensitive.

The rice plant (like all other plants) has a number of signal mechanisms which initiate each stage in its life sequence. The signal mechanism usually takes the form of a substance which chemically passes a message to initiate a new stage of growth. The "trigger" which initiates the signal for flowering in many species is the length of the nights in late fall. Such species are said to be photoperiod sensitive or photosensitive. Some but not all varieties of rice fall into this category. The genetic basis for this distinction is not known. If the distinction is determined at a single gene locus, there may be some problem in explaining the existence of "weakly photosensitive" varieties (Tanaka, et al., 1964).

Photosensitive varieties seem particularly well suited to cultivation during the tropical monsoon season. Non-photosensitive varieties are well-adapted to conditions where a short growing season is a limiting factor. This is the case with Sansai farmers who work fields of "medium" altitude, and attempt to raise two annual rice crops, and a third crop of garlic or soybeans.

Factors Controlling Plant Growth and Yield

The key process in the growth of the rice plant (or any other plant) is photosynthesis. This is a process in which water, oxygen, and carbon, and small quantities of other substances, are processed to form organic substances, consisting mainly of carbon and hydrogen combined in various ways. Roots, leaves, stems, flower, seeds hormones, amino acids, and starches are synthesized from water, oxygen, and carbon.

The CO_2 used by the plant is derived from the atmosphere (although some may also be derived from water-saturated soil) which consistently contains the gas in the proportion of three parts in 10,000. This factor must be taken as a "given," and not as an important variable. There are a number of variable factors which affect the growth patterns of the rice plant. *Sunlight* provides the energy necessary to carry on the operation of photosynthesis. If sunlight is lacking, either in quantity or intensity, growth of the plant may be inhibited. *Water*, provided by rainfall and irrigation, is of considerable importance in the growth of the plant. *Temperature* determines the rate at which essential biochemical processes take place, and extreme highs or lows may result in complete cessation of such processes. The availability of certain *nutrients* which are present in the soil, in irrigation water, and in fertilizer, is important to plant growth.

In considering the physiological requirements of the rice plant, the factor of time cannot be ignored. The plant has different needs at different times or stages of growth. Appropriate temperatures for one stage of plant growth may be unsuited for another stage. More water may be needed at one time than another. Long periods of sunlight may be essential at one stage, and not particularly important at another.

Because rice is a crop plant, raised for the food, one aspect of growth is all-important. This is the growth pattern which leads to production of mature seeds, or grain. Conditions which encourage rapid vegetative growth do not necessarily result in production of large amounts of grain. Discussion here centers mainly on conditions and agricultural methods that do lead to maximum yield of grain.

There are numerous relationships between the conditioning factors. For instance, with increased sunlight and heat, the transpiration rate increases, and greater amounts of water are needed. The temperature of water standing on wet rice fields has a moderating effect on air temperature.

Grain yield depends primarily on the amount of plant material formed during the ripening stage late in the season. Although some carbohydrates are translocated from stems and leaves formed earlier in the season, the great bulk of grain carbohydrates are transferred directly to the panicle after assimilation during the last few weeks of the season (Tanaka, et al., 1966:6). Grain yield depends largely on the quantity of photosynthetic product during ripening.

The energy requirement of the rice plant varies during the season, but remains substantially below the level provided by full tropical sunlight--approximately 700 calories/cm^2/day. In the early stages of growth of virtually all leafy, tropical plants, dry matter can be produced when the solar radiation rate is only 100 calories/cm^2/day. Later, during the flowering stage, dry matter production requires inputs in excess of 350 calories/cm^2/day.

Full tropical sunlight is more than adequate to support photosynthesis. However, photosynthesis is not continuous, while respiration is. The former ceases during darkness, and may be slowed on days which are heavily overcast. Heavily shaded leaves often cease to photosynthesize. Meanwhile, the process of respiration continues but at considerable cost in dry matter. Darkness, cloudiness, self-shading and mutual-shading may cause some leaves to become nonproductive. For example, such leaves expend greater energy in respiration than they fix through photosynthesis. During the period of panicle formation in the rice plant, heavily shaded leaves may inhibit, rather than enhance, grain formation. Large, leafy, "tropical" rice plants may produce relatively little grain. Experiments with plants of this type sometimes have shown a negative response to nitrogen fertilization. The most productive rice plants in the world are the relatively small, spindly varieties of Japan (Tanaka, et al. 1966).

Sansai farmers are able to specify, at least casually, the potential relationship between heavy leaf production, shading, and light grain production. Several residents reported the experience several years ago of a local household which applied fertilizer to a local variety. The experiment resulted in rice plants that were strong, leafy, and "pretty." But grain production was unusually low. Sansai farmers perceive the following causal sequence--nitrogen fertilization = "pretty" plants = low yield. They regard this sequence as applicable to rice varieties in general. It is probable, however, that the rule is of limited application, providing a useful guide only to cultivation of tall, leafy, "tropical" varieties.

Within wide limits, temperature does not appear to affect the photosynthesis rate of the rice plant. Solar radiation is the major determining factor in this process (Tanaka, et al. 1966). However, the respiratory rate of the plant is importantly affected by temperature. As temperatures rise, dry matter loss to respiration rises. At the same time dry matter production through photosynthesis remains at the same level. The result is lowered efficiency in grain production. To achieve maximum yield, ample sunlight should be combined with moderate temperatures during the ripening period. The final weeks of the late rice crop at Sansai--during November and December--fulfill these ideal conditions moderately well. The early rice crop is harvested during the monsoon rains when sunlight is sparse and temperatures are quite high.

Since the evapotranspiration rate of the rice plant is affected by sunlight and temperature, the water requirements cannot be stated with precision. It can be assumed that fields which have *some* standing water throughout the period preceding ripening are adequately watered. In most areas this condition is guaranteed by a seasonal water supply of about 40-60 inches.

Too much water is no less a matter for concern than too little. Because of unpredictable rains, rice farmers are inclined to retain all water that is captured on the fields, even if the height is thought to be excessive for optimal growth conditions. Little attention ordinarily is paid to drainage systems. If heavy rains come, there may be broken bunds, deep water, and drowned plants. Sansai farmers face problems with too little water during the early rice crop, and too much during the late crop.

Regulating Biological Competition

The rice plant must compete with other species for nutrients, water, CO_2, sunlight, and space. The purpose of the farmer is to alter the competition in favor of the rice plant. But the farmer is not completely successful for the rice plant must compete with microorganisms that grow in, on, or around it (diseases), weeds, and numbers of small animals and insects (pests).

Grist reports that "About forty diseases attack paddy in the field and of these the most prevalent and important are 'blast' (*Piricularia oryzae Cav.)*, foot rot or 'bakanae' disease (*Gibberella Fujikuroi)* and seedling blight (mainly *Helminthosporium oryzae)*" (Grist 1959:265). "Blast" disease attacks the plant leaves, culms, branches of the panicle, and floral structures. The condition symptomatic of the worst possible attack is "rotten" neck in which lesions appear on the neck of the culm and on the lower branches of the panicle. This disease can be carried by infected seed. If possible, farmers should avoid using seed from infected stands of plants. Varieties vary considerably in their susceptibility to this disease (Grist 1959:267). Paddy raised in a wet nursery (with standing water) is more resistant to "blast" than paddy raised in a dry nursery. Use of green manures favors development of the disease.

One prominent symptom of foot rot is that leaves dry up one after another from the bottom of the plant. Eventually, there is evidence of fungus attack at the lowest (below ground) nodes and internodes (Grist 1959:276). The disease may be carried over from one season to the next by contaminated seed or in the soil. The organism survives best in poorly drained soil.

Symptoms of seedling blight include darkness at the base of the stems and roots, a stunted appearance, and the leaves which wither and turn yellow. Affected plants die slowly or they may recover when the fields are flooded. The fungi grow best in soils that are rich in organic materials, e.g. manure, night soil, etc. The disease can be spread by flowing irrigation water. The organism thrives in warm, moist weather, and the disease is most severe in dry nurseries.

Wet rice fields are flooded during part of the year, but usually dry during several months. The result is an alternation between aquatic and dry land weeds. Weeds of the dry land type ordinarily are plowed under before the start of the growing season. If these plants re-emerge, they may provide some competition to sown rice at first, because the new plants must be given some time to establish themselves before the fields are flooded. If the rice is transplanted into the fields, then similar weed problems will be encountered in the nursery.

During most of the growing season, when fields are flooded, weeds of aquatic and semi-aquatic type are a problem. Semi-aquatic weeds that frequently raise problems for rice farmers include the genera *Cyperus* (sedges), *Typha* (cattails), and *Echinichloa* or water grasses (Grist 1959: 223). *Cyperus* species may be either annual or perennial, while *Typha* species are perennials, and *Echinichloa* species are annuals. All flourish where land out of cultivation is poorly drained. Good drainage in the off-season will help to eliminate these weeds, but will improve conditions for dry land types.

Insects are the most important pests of the developing rice plant. Hundreds of insect pests are known, but only a few are consistently harmful in a given area. The worst pests, some representatives of which are found in every rice growing area, include plant bugs, stem borers, army worms, and grasshoppers (Grist 1959:236). A number of individual pests will be discussed as examples of the problems that rice farmers face.

Schoenobius incertulas (Wlk.) or the "yellow stem borer" is the larva of a moth. The parents live an average life of five to seven days, are nocturnal, spend the day hiding under leaf sheaths, and mate in the late evening. Eggs are laid in clusters of thirty on the under sides of leaves. After a six or seven day incubation period the newly hatched larva emerges, and makes its way down to bore into the stems. In addition to crawling, the larva may also be transported great distances to other rice plants by wind and irrigation water. Only about ten to twenty-five percent of the larvae succeed in finding a host. The larva eats the inner tissue of one stem, and then moves to another. Each larva is capable of injuring two or three stems during its life. Larva eventually crawl down a stem and pupate near the base. This stage lasts six to fourteen days, after which the moth emerges. On irrigated fields, they can emerge from four or five inches of water. In some cases the larva hibernate, and may last through the winter if stubble is left on a dry field after harvest (Grist 1959:240-242).

Spodoptera mauritia is a leaf or "army" caterpillar which appears when paddy seed is germinating. Young plants in wet nurseries are selected for egg laying. The eggs are laid by moths that characteristically appear in large numbers. The eggs hatch in four or five days, and the caterpillar which emerges crawls from plant to plant, feeding at night. The caterpillar feeds for about four weeks before it pupates, and emerges as a moth. Attacks by these caterpillars usually are so sudden that the cultivator is unaware of the danger before his crop is destroyed. The most effective control measure is flooding the nursery.

Hispa armigera (Oliv.) is a leaf beetle. Its eggs are laid on a variety of grasses, including cultivated rice. The young larvae, upon emergence, "mine" leaves by eating the green matter and causing the leaves to wither. The larvae emerge from the leaves in the afternoon, and crawl to other leaves. Considerable damage can be done (Grist 1959:257).

Other important pests of growing rice include crabs, birds, and rats. Both sea and land crabs may enter paddy fields and cause damage by feeding on young plants, burrowing into the retainer walls and interfering with water control. Land crabs are a greater problem for they do not migrate. Traps may be used to capture them. In Sansai children capture them by hand, and they constitute an occasional source of food. Birds can be a problem at all stages of growth, but especially when the grain has formed. They are a particular problem where there is plenty of cover adjacent to the fields. The most effective means of their control is scaring by a variety of methods. Rats are destructive of rice plants at all stages of growth. Control by poisons or traps is difficult since rats are suspicious animals (Grist 1959:260-263).

Decisions "Imposed" by the Rice Plant

Rice cultivation imposes innumerable decisions on the farmer. Many of these, perhaps most, affect potential yield and profit. Decisions are interrelated in complex, unpredictable ways. In selecting a variety for planting, the farmer may consider, among others, the following variables: glutinous vs. nonglutinous type; photosensitive vs. non-photosensitive type; tall, leafy vs. short, spindly type; resistant to drowning vs. non-resistant; resistant to dessication vs. non-resistant; resistant to "blast" disease vs. non-resistant; and positive response to nitrogen fertilization vs. negative response. Any variety is likely to manifest some positive and negative characteristics and some which are neutral in a given environment.

In fact, most choices made by the farmer have potential effects which are both positive and negative. Application of manure generally improves plant growth. This often is associated with improved yield but in some cases "pretty" plants are associated with lower yield. In addition, the manure also provides nutrients to weeds and disease organisms. Raising seedlings in a dry nursery provides a degree of resistance to desiccation, but lessens resistance to "blast" disease.

The farmer exercises choice over many variables in the system, but accepts others as given. He makes significant choices regarding glutinous versus nonglutinous types, photosensitive versus non-photosensitive types, and specific variety. Individually, he has little control over water supply. It may be rain-fed, a relatively inefficient irrigation system, or a massive government-operated scheme. He does exercise choice over use or nonuse of tractors, insecticides, and fertilizers. If he opts against these innovations, it is assumed he does so on the basis of sound economic reasoning, not on the basis of "peasant conservatism" or "tradition."

Early Rice Cropping Pattern

At present the early rice crop at Sansai is planted exclusively in nonglutinous varieties. These varieties are not photosensitive. They can be raised during any period of the year in which water resources, heat, and sunlight are sufficient. In Sansai a few households set rice seed in nurseries as early as February 2, and others as late as March 25. However, nearly seventy percent of village households sowed their nurseries in the much shorter period between February 11 and February 28. The "mean" planting date was February 17.

Early rice seedlings were transplanted in 1970 between March 3 and April 25. Sixty-three percent of households surveyed began transplanting between March 20 and April 10. The "mean" date for initiating transplanting was March 30. Typically, seedlings were transplanted about six weeks after the nursery was sown.

Some Sansai households began harvesting the early crop of 1970 as early as June 6, others began as late as August 10. The "mean" date for initiation of harvest was July 8. Sixty-four percent of the households surveyed began harvest between June 26 and July 20. The "typical" growing season for early rice in Sansai extends from February 17 to July 8, a total of 141 days. The period between transplanting and harvest extends for 100 days--March 30 to July 8.

Agricultural Tasks During the Early Rice Crop

During the early rice crop the following major tasks must be carried out: preparation of nursery bed; watering nursery; plowing/harrowing; "cleaning" the field; pulling, cutting, and carrying the seedlings; transplanting; weeding; irrigating the fields; reaping; threshing; and carrying paddy to storage. In addition to these major tasks, there are a variety of minor tasks. These tasks include: inspection and repair of the bunds; application of insecticide; selling of surplus rice; and making rental arrangements. These major and minor tasks are now discussed in greater detail.

Preparation of Nursery Bed

Rice seedlings in Sansai are raised in the semi-wet type of nursery described by Watabe in *Glutinous Rice in Thailand* (1967).

> The nursery beds seen in the Chiang Mai Valley are the so-called semi-wet or semi-irrigated types. The bed is a strip of land about 1 to 1.5 meters in width, rectangular in form, and raised to a height of about 10 to 20 cm. Between adjoining beds, a narrow ditch-like strip about 30 cm. wide is left for use as a foot path, and for irrigation purposes. The foot path section is then dug out and the dirt is piled up on either side to form ridges. The surface of the nursery bed is worked over with hoes or sticks to soften it completely, and then it is finished off level with the flat of a bamboo stick or a strip of wood (1967:60).

In some cases, the nursery is prepared only with a hoe; no draft animal or plow is used. Once the nursery is prepared, the task of broadcasting the seed occupies only a few minutes. In February and March, the semi-wet nursery must ordinarily be prepared in the lowest *haj myang* (walled plot) of available fields. If the nursery is too high, irrigation water will be unavailable.

Watering the Nursery

The semi-wet nursery has the advantage of adaptability to varying climatic conditions and water supply. If the water supply is adequate but not excessive, the small ditches will be nearly full, and the seedlings receive sufficient moisture through seepage. If there is only a little water in the small ditches, a woven scoop is used to transfer water from the ditch to seedlings. If there is no water in the ditches, watering cans carry water from the nearest irrigation ditch, canal, or well. If the water supply is "excessive," it will only convert a semi-wet nursery to a wet nursery. As long as the seedlings extend above the water surface, there is no problem.

"Watering" takes several different forms. It may involve the use of large watering cans, woven scoops, and well or canal water. This method is used also for watermelons and garlic. It may require the temporary breaching a bund, allowing water to flow directly from irrigation ditch to nursery, or through a series of plots to the one where the nursery is located. If necessary, the irrigation ditch can be dammed temporarily to force water into the appropriate field. If water is unusually high early in the season, the farmer may instead be concerned with drainage. No fertilizer, natural or chemical, is applied to nurseries in Sansai. This is in line with Watabe's observations throughout the Chiang Mai plain and northern Thailand (1967:63).

Plowing/harrowing

Sansai farmers consider these two tasks as part of a single operation. Although they were asked to give separate information on labor expended for plowing and harrowing of the early crop, many farmers were unable to do so. Watabe provides the following description for all of northern Thailand. It is generally accurate for Sansai:

> The first operation in a paddy field is the plowing, which takes place from two to seven days before transplanting. For this purpose, the native wooden plow, which is a kind of Chinese plow, is used. The sole of the plow is about 50 cm. [square?] and can be easily maneuvered, but it is not possible to plow very deeply. . . . The coulter is made of iron. . . . Generally, the plow is pulled by a buffalo or an ox, but occasionally by two of them together. The plow is single furrowed and always turns to the left (1967:65).

Only a few oxen are owned in Sansai and they are seldom used for plowing or harrowing. Sansai farmers prefer to plow already flooded fields where possible. Water buffaloes are more effective draft animals than oxen in deep mud.

Watabe offers the following description of "puddling" or harrowing:

> Puddling is carried out on the same day as the transplanting. Puddling is also done with either a buffalo or an ox, using the native harrow. They work both crosswise and lengthwise across the paddy fields filled with water. The harrows are made of wood, but bamboo, wood or iron is used for the teeth, which are about 12 cm. long, and places about 13 cm. apart. Right after puddling is finished, leveling or smoothing is started. For this operation, they have either a buffalo or an ox pull the harrow, which now has a bamboo pole about two meters long and a diameter of about 15 cm. added across the teeth section (1967:65-66).

In Sansai, harrowing sometimes is done a day or two before transplanting is begun. This is so the field can be thoroughly "cleaned" before the next step. The "leveling" operation often is omitted because "cleaning" has much the same effect.

"Cleaning" the Field

If the field is not deeply flooded for plowing and harrowing, visible grasses and weeds are ordinarily pulled by hand. But the most common method is for women and children to "tread" the weeds a few inches below the surface of the mud. Since this task is done in muddy water, weeds are located by touching them with their feet.

Pulling, Cutting, and Carrying the Seedlings

Before the seedlings are transplanted, they must be removed from the nursery. Watabe's general description applies well to Sansai practice. Usually the seedlings are pulled up one or two days prior to transplanting:

> Grabbing the plants around the tops, they pull up at one time, a bunch consisting of about 20 to 30 plants. Shaking off the excess mud sticking onto the roots by beating the bunch against the soles of their feet, they take 70 or 80 plants to make a bundle and tie them up with thin bamboo strips. Since by this time, the seedlings have generally reached a height of from 60 to 70 cm., the tops are chopped off, leaving seedlings of an equal length of about 40 cm. There are at least three good reasons for this cutting process: (1) making the labor of transplanting easier. (2)

> lessening the amount of evaporation from the seedling, and (3) limiting the excess vegetative growth (1967:66-67).

Watabe's last point is related to the fact that vegetative growth can sometimes be excessive. A plant with too many tillers and leaves may have reduced grain yield because of excessive self-shading in the important period after flowering.

Transplanting

After the field is fully prepared, so that it consists of smooth, level mud covered by two to four inches of water, transplanting begins. Once again, Watabe's general description of northern Thai practice applies to Sansai:

> In the methods of transplanting commonly employed here, the planters transplant stepping backwards, and it is on rare occasions that they move forward. The planters, holding bundles of seedlings in one hand, transplant four to six plants to a hill. According to the writer's observation, recorded at 34 paddy fields, the spacing between hills is about 32 cm., that is about 9.9 hills per square meter. This average figure is almost half that of the Japanese (1967:67).

In the two cases where distance between rows was measured at Sansai, the figures calculated were 31 and 33 cm. Watabe presents data suggesting that for alluvial, double-cropped fields of the type found in Sansai somewhat closer spacing would result in higher yields (1967:67).

Weeding

There is substantial variation in the amount of labor devoted to weeding in Sansai. Watabe suggests that weeding is carried out "once or twice within the month after transplanting" (1967:70). In fact, some Sansai households do little or no weeding, while others carry out several successive weeding operations. Weeds ordinarily are pulled by hand. Watabe writes that:

> In observing throughout the whole of northern Thailand, one gets an impression that weeds are not particularly abundant in the paddy fields. Several reasons for this can be cited. The first one is that in the deeply irrigated paddy fields, only aquatic weeds can exist. The second one is that the seed or subterranean roots are destroyed by the dryness and high temperature that exists during the hot season. The third one is that after harvest, the burning of the straw in the fields has the same effect of cooking out the seed and root systems of weeds (1967:70-71).

While fields are seldom burned in Sansai, the practice of raising two or three crops on the same plot each year is probably an important factor in weed control.

The most common weeds in the rice fields of northern Thailand are *Cyperus* species, *Fuirena glomerata, Fimbristylis miliacea,* all members of the sedge family, and *Digitaria marginata, Eleusine indica, Leersia hexandra,* all members of the Grass family (Watabe 1967:71). No herbicides are used in Sansai.

Irrigation

The highest fields of the village are not used for early rice. In March and April there is little local rain so only relatively low fields can be irrigated soon enough to facilitate plowing, harrowing, and transplanting. Usually, at this season, local irrigation headmen must go to the Irrigation Department office near the dam at Don Kaew to request that the sluice gates for *Myang Çaaw Mahawong* be opened. On the day designated, farmers are able to irrigate by temporarily breeching the bunds surrounding their plots, and temporarily damming the irrigation ditch, if necessary.

In recent years, gasoline-operated, irrigation pumps have been introduced in Sansai. Their effect is to improve considerably the reliability of the water supply for the early rice crop. About 50 *rai* of rice fields in the southern portion of Sansai were irrigated by six pumps in 1970. Each pump services the plots of three to ten households. Each pump is located at the highest spot among the fields to be irrigated. A pipe is driven as much as five to six meters into the ground to obtain water. In some cases, water is directed across irrigation ditches in rubber hoses or bamboo tubes. The cost of pump irrigation is ten buckets of paddy (100 kg.) per *rai.*

Reaping

Mature plants are reaped with a short, curved sickle. The heads of a number of plants are grasped with one hand, and the stalks are cut about 25-30 centimeters above the ground. Ideally, the cut stalks are laid on the stubble and allowed to dry in the sun for two or three days (Watabe 1967:72). Since the harvest of the early rice crop is carried out in the middle of the rainy season, water and moisture cause problems in harvesting and storing the crop. The writer's journal for July 25, 1970 includes the following insights into harvest problems at this time of year:

> The last few days, people have been carrying unthreshed rice from the fields on *khaa maa* ["dog's leg"--a type of wood and bamboo carrying rack] because the fields have standing water. The rice must be threshed in a high, dry place. Cut rice is laid on the stubble, (hopefully) above the level of the water. People cut rice in the morning, and carry it on *khaa maa* in the afternoon. Usually, there is not so much of a problem. This year it rained more than usual so there is already an excess of water.
>
> The cooperative labor groups are unusually large this year. Often groups of 40-50 work together, instead of the usual five or ten. This is an apparent response to greater problems involved in the harvest this year. A few households have been forced to harvest from dugout boats.
>
> Some people have lost rice through water damage. *Naaj Wanlop* cut and stacked his rice (in a circular pile, head in) as Sansai villagers often do in dry years for late rice and occasionally

for early rice. He planned to leave the threshing until later. He then went off to buy *longans*. The water in the field rose rather quickly, and he lost perhaps one-half of the crop.

Threshing

After reaping, the bundles of rice stalks are dried and carried to the threshing place. The thresher uses a simple tool consisting of two lengths of bamboo, each about two feet long, connected by a cord of the same length. The cord is wrapped around the bundle of stalks in such a way that the bundle is tightly gripped, and the bamboo lengths lie side by side. This device effectively lengthens the "striking arc" of the thresher. He then beats the heads of the rice stalks against the inside of the *khu* (harvest basket). The harvest basket is 2.0-2.5 meters in diameter. Seven to ten strokes are sufficient to remove virtually all grain from the stalks. The grain is then winnowed by throwing it out of the basket in a high arc, onto a woven mat, using a wooden shovel. Chaff is removed by creating an artificial breeze with a pair of large, woven fans.

Carrying the Rice to Storage

The threshed rice is then carried to the storage bin. In Sansai, ordinarily this is done by human labor. Because of the short distances and the expense involved, oxcarts are seldom used. The carrying device used (*haab*) consists of a flexible, split-bamboo pole, and two suspended carrying baskets. Typically, the baskets are loaded to within two or three centimeters of the top. In this case, a load of about thirty kilograms of paddy can be carried. The weight of the entire load is about thirty-two kilograms.

Before the rice can be stored, often it is necessary to re-dry it. If this is not done, particularly for the early crop, the grain will have poor milling qualities. It may also be attacked by several species of fungi. The writer's journal entry of July 25, 1970 provides this information:

> People are having trouble drying rice, even after it has been threshed and carried in from the fields. They spread it very thin on mats, and stir it around from time to time. They can put it under the house or out under the sky. But it often sprinkles these days. The longer it takes to dry, the more they will lose to chickens, etc.

Minor Tasks

Since most Sansai fields produce two or three crops annually, the repair and maintenance of bunds is a year-around task. The amount of time devoted to this job in any one season is small. Several other tasks are carried out only by a fraction of local farmers, and none require much time. Renters ordinarily spend an hour or two each year reaffirming their rental arrangement. At issue, in particular, is the division of *kaa kaan* (provision of labor for the irrigation system and related projects) and *kaa khwaaz* (provision of a buffalo or ox for plowing and harrowing). The landlord ordinarily provides either the labor or the animal and the renter the other. Rented rice land is always sharecropped on a 50/50 basis. The few farmers fortunate to produce rice in excess of their household needs devote a few hours to arrangements for its sale. They observe the measurement of the paddy as it is transferred from the rice bin to the dealer's truck.

Rice Varieties Planted During the Early Season

The primary problem that the Sansai farmer faces in raising the early rice crop is inadequate irrigation water at the start of the season. Therefore, each farmer attempts to plant a variety which combines the qualities of resistance to desiccation with high yield. Theoretically, there are a number of varieties to choose. Various informants reported that at least six different varieties have been used in Sansai in recent years. (Maps 6 through 8 demonstrate shifts in varietal choice between 1968 and 1970.) However, during the early rice season of 1970, only two varieties were planted in significant quantity. Locally, these varieties are called *khaaw lyang nauj* (yellow, small-grained rice) and *khaaw kaseet* (rice of the Ministry of Agriculture).

Before 1969, *khaaw lyang nauj* was well-established as the preferred variety in Sansai. It had a satisfactory resistance to desiccation, and a reasonably high yield. However, in 1969 one farmer adopted a new variety under the following circumstances:

> Mr. Dii is a resident of Longphumoon (a nearby village) who farms a plot of land lying just within Sansai's southwestern border. In 1969 he got seedlings from his younger brother who lives at Baan Saan in Lamphun Province. His brother had received and planted the seeds only a matter of weeks before. Mr. Dii thinks that his brother received the seed in a plastic bag from a local district official.

By 1970, six additional households were planting *khaaw kaseet* in Sansai. Many more households planned to do so in 1971. Farmers obtained seed of a new variety by trading an equal amount of some other variety to a farmer already raising the improved type.

The introduction of the new variety into Baan Saan, Lamphun, was a project of the Rice Experiment Station at San Pratong. The rapid acceptance of the new variety demonstrates the success government programs for agricultural innovation can enjoy when the new product is clearly superior to its predecessors.

At the Rice Experiment Station, the new variety is called *Kau Khau Saam*. It is a cross between *khaaw lyang nauj* (already used in Sansai) and IR-8 from the International Rice Research Institute at Los Banos, Philippines. This variety can be planted at any season, i.e. it is not photoperiod sensitive. Other specifications under local conditions for *Kau Khau Saam* are as follows: maturation period, 120-130 days; height, 90-100 cm.; resistant to "lodging"; brown husks; resistant to disease; and expected yields, 800-1,000 kg./*rai*.

Mean yield figures for *khaaw kaseet* are somewhat higher than those for *khaaw lyang nauj*. (The main reason that *khaaw kaseet* yields are lower than predicted by government agronomists is that Sansai farmers use no chemical fertilizer in rice cultivation.) Forty-nine households averaged 488 kg./*rai* with the latter, while six households averaged 575 kg./*rai* with the former. However, the difference between the two means is significant at the .02 level.

Map 6 *Early Rice Crop Choice at Sansai, 1968*

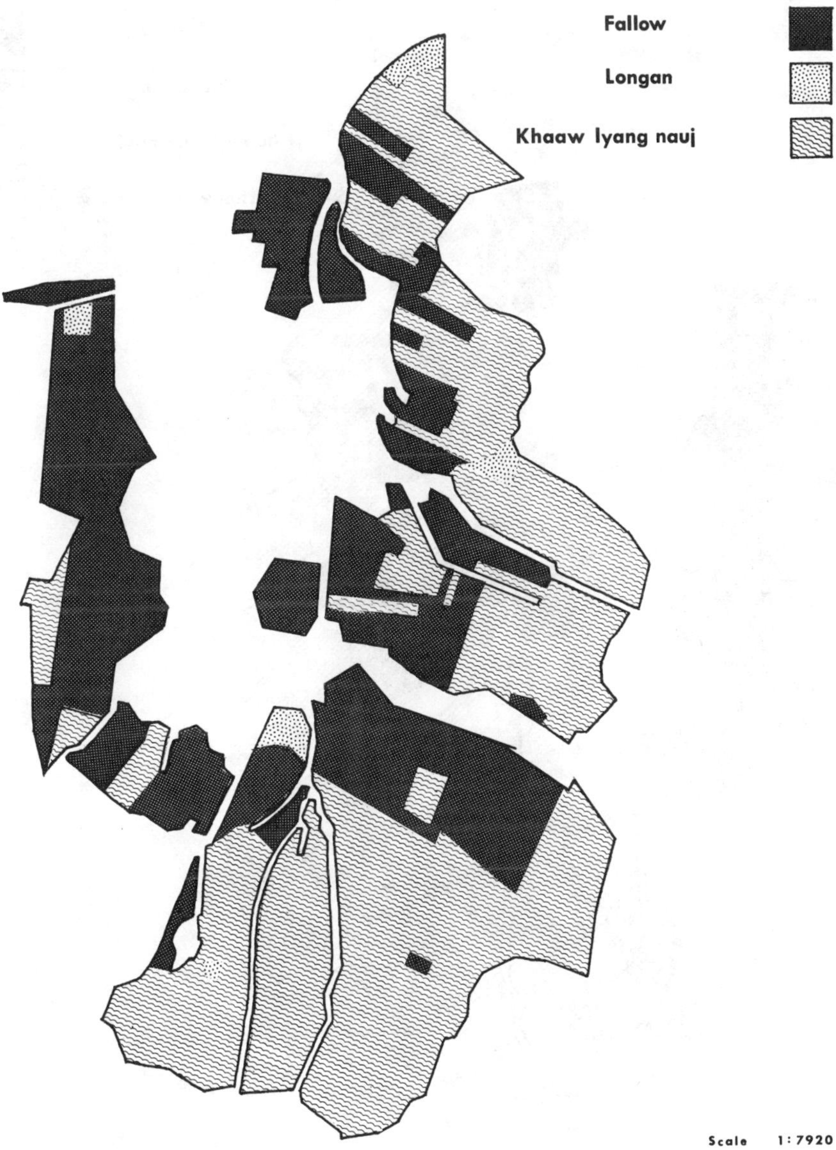

Map 7 *Early Rice Crop Choice at Sansai, 1969*

Map 8 *Early Rice Crop Choice at Sansai, 1970*

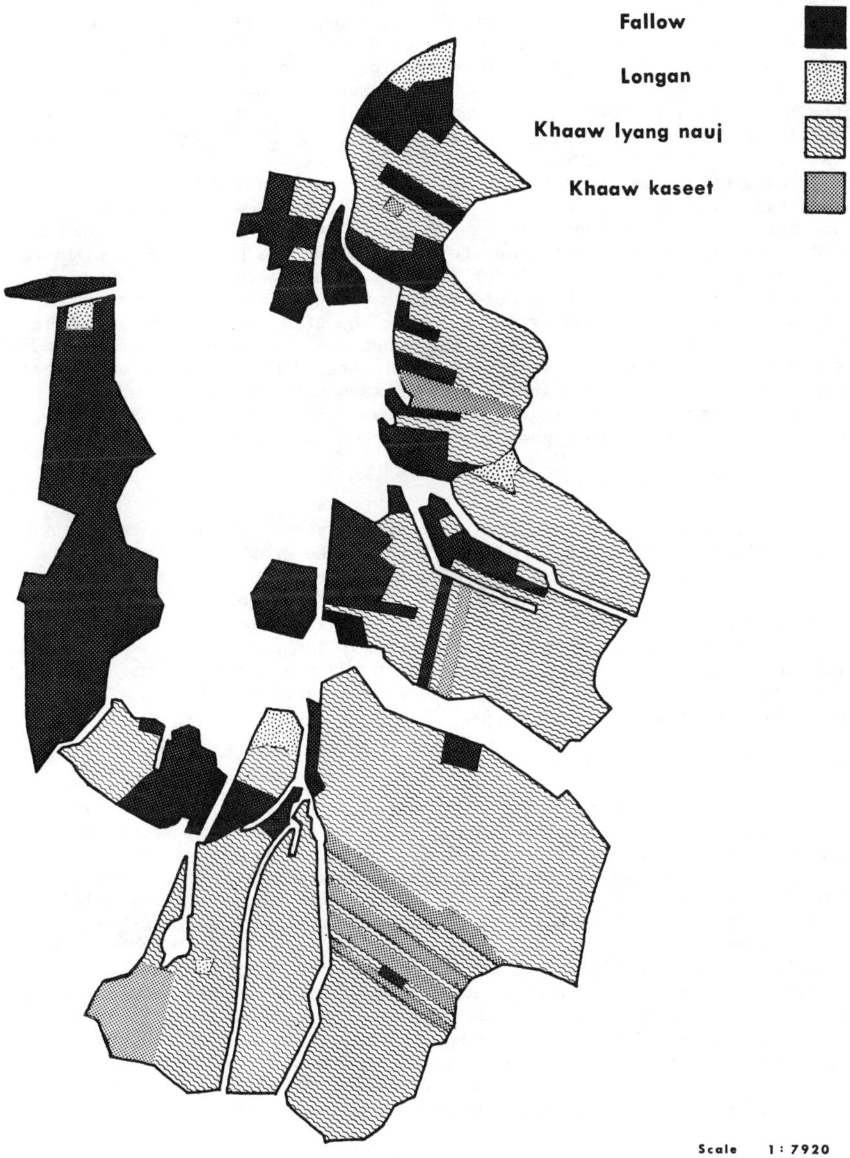

Paddy yields are not the only basis upon which varieties can be compared. Milling ratio--the percentage of paddy weight which is converted into edible rice--it also significant. Measurements were carried out at both local rice mills in July, 1970. Measurements of twenty-nine separate samples of *khaaw lyang nauj* produced variations in milling ratio from 51.2 to 67.7 percent. The mean figure was 61.8 percent. Measurement of four samples of *khaaw kaseet* produced a range of milling ratios from 51.6 to 61.4 percent. The mean figure was 58.3 percent. The data suggest that the higher yields provided by *khaaw kaseet* are slightly offset by lower milling ratio. Poor milling quality has been one of the primary criticisms of the IR-8 variety.

Edible rice is not the only "yield" of Sansai paddy fields. Another important produce is rice straw. Sometimes it is used for buffalo fodder. But its major use in Sansai is in the garlic gardens planted in November and December. After the fields are prepared and the garlic cloves planted, the beds are covered with a layer (roughly five cm. thick) of rice straw. The straw acts as a mulch, providing both organic matter and protection against weeds. In November and December the straw has substantial cash value; there is a lively trade in the substance. Farmers unable to obtain straw from their own fields or by purchase are forced to raise some other garden crop. During the garlic planting season, large trucks loaded with straw often are seen passing through Sansai from north (where no garlic is raised) to south (where a great deal is grown in Lamphun Province).

Late Rice Cropping Pattern

In essentials, the late rice crop is not very different from the early crop. Since the tasks required in producing the crop differ little from those described earlier, the list will not be repeated. Labor inputs per *rai* vary little between the two crops. Fields used for late rice, like those used for the early crop, are small and within easy walking distance of the village. Both crops provide straw for the garlic beds and fodder. Yet there are several particulars in which the two crops differ. Most important among these are: cropping schedule; varieties planted; and major problems encountered.

Late Rice Cropping Schedule

The late rice crop of 1969 was planted over a long period of time. One village household planted their nursery as early as May 20, while another did not plant theirs until September 27. Within a single village "starting" dates vary as much as 130 days. Households that planted very early (before July 10) were those who own or rent land too high for successful cultivation of the early crop. Households that planted very late were those with low land who anticipated a flood in late August or September. Most of these households waited until the floods of late August began to recede in early September. Between August 4 and August 22, no rice was planted in nurseries in Sansai. Some 51 households (54.3 percent of those surveyed) planted their nurseries between May 20 and August 4. An additional 43 households (45.7 percent) planted between August 22 and September 27. The "average" planting date for the former group was June 25, and for the latter, September 13. The late rice crop is transplanted a few weeks after the nursery planting.

Households that planted early (May 20 to August 4) transplanted between June 24 and August 25. The "average" date for transplanting was August 2, and 37 days, on the average passed between planted and transplanting. Households that planted during or after the late August flood transplanted between October 3 and November 26. The "average" date for transplanting was November 14, and the typical period between planting and initiation of transplanting was 62 days.

Those Sansai households that planted the late rice crop early (May 20 to August 4) began harvesting between November 7 and December 20. The great majority of these households began harvesting in the period between November 20 and December 10. "On the average" this group began harvesting on December 3. The average growing season for late rice planted before August 5 was 162 days.

However, this "average" figure conceals a wide range of values for growing season length. For example, the household that planted on August 4 began to harvest on December 12. Their rice required only 131 days to mature. On the other hand, the household which planted on May 20 did not begin to harvest until November 10. Their rice took 174 days to mature. This sizable variation is explained by the fact that all glutinous rice varieties raised in Sansai are photoperiod sensitive. After the plant has reached full vegetative growth it enters a lag phase which lasts until the short day lengths of late fall signal to the plant that the time has come to initiate the period of flowering and maturity. Glutinous varieties planted in Sansai have a potential growth period varying from less than 120 to more than 200 days.

Rice which is planted too late does not reach full vegetative growth before the day length signal is given in the late fall. In this case, the plants take two or three additional months to mature. Sansai households which planted after August 21 began to harvest between February 2 and March 11. The "average" date for initiation of harvest was February 17. Households that planted after August 21 had to wait an average 157 days before their rice matured. Perhaps it matured when it did because day lengths were again the same as they had been in late November.

Sansai households, as demonstrated, choose between two distinct schedule options to cope with late summer floods. A number of households try to plant and transplant before the floods. Others choose to "wait and see," planting *after* the floods, *if* they come soon enough. A number of households were forced to attempt both alternatives in 1969. A few households planted only a few days before the flood. They saw their very young seedlings drown before they were well-established. These households were forced to replant after the waters receded. They became part of the late planting group discussed earlier.

A few of the households flooded out gave up on the 1969 late rice crop, and chose to wait until November to plant garlic. Those households that were successful in planting early, planted between May 20 and August 4, began transplanting between June 24 and August 25, and began to harvest between November 7 and December 20. Those who chose to plant after the flood water receded, or were forced to replant at that time, planted between August 22 and September 27, began transplanting between October 3 and November26, and began to harvest between February 2 and March 11. In one household or another, the late rice crop of 1969 occupied Sansai residents for more than nine months.

Rice Varieties Planted During the Late Season

Sansai residents planted three different glutinous rice varieties for the late crop in 1969. They were: *khaaw kaew dau* (old, early rice); *khaaw kam phaaj;* and *khaaw lyang San Pratong* (yellowish rice from the San Pratong Experiment Station). (Maps 9 through 11 illustrate the shifts in varietal choice between 1967 and 1969.) Of Sansai households surveyed, 38.3 percent cultivated *khaaw kaew dau,* 31.9 percent cultivated *khaaw kam phaaj,* 5.3 percent raised *khaaw lyang San Pratong* and 24.5 percent raised a combination of two or three varieties. In choosing a variety suitable for the late crop, Sansai farmers seek two major qualities: early maturation date and resistance to drowning. No variety presently available at Sansai receives a high rating for both these variables. Secondary considerations include milling ratio and eating quality.

As recently as 1968, other rice varieties were raised at Sansai. Two varieties were *khaaw Singapore* and *Khaaw pha lyat.* In 1969, and in their plans for 1970, Sansai farmers had settled down to the three varieties mentioned earlier. In explaining their choices and reasons for changing from one to another, local farmers provided the following insights into significant qualities of the three currently used varieties.

First, according to information provided by local residents, *khaaw kaew dau* was introduced in Sansai around 1960. Its chief attraction to local farmers is its early maturation date. When sown by early August, the usual date to begin the harvest is middle or late November. This is particularly significant to farmers who plan to plant garlic. It is widely believed that garlic planted as early as November, and therefore harvested early, will receive a price higher than usual. *Khaaw kaew dau* is said to be moderately resistant to drowning, but has poor resistance to disease. Several farmers mentioned that their fields of *khaaw kaew dau* had "yellow leaf disease" in 1968 and 1969.

Several farmers who have changed to *khaaw kam phaaj,* or are planning to do so, indicate that their reasons centered around the superior resistance of the newer variety to drowning and disease. The few farmers who indicate a change from *khaaw kam phaaj* to *khaaw kaew dau* express an intention to plant garlic, and therefore are attracted by the early maturation potential of the latter.

Second, information from Sansai farmers suggests that *khaaw kam phaaj* has been cultivated in the village since 1965 or 1966. It has several qualities to recommend it. It is resistant to drowning and disease. Because it has a strong stock, it is also resistant to "lodging." Because of these qualities, it is particularly well-suited to use in Sansai's lowest fields. Many farmers indicate changes in the recent past or near future to this variety from *khaaw kaew dau* or *khaaw lyang San Pratong.*

Third, *khaaw lyang San Pratong* was introduced into Sansai in the mid-1960's. It is suitable for cultivation only in Sansai's highest fields. Where irrigation water is adequate, but not excessive, and all other conditions are nearly ideal, it is capable of producing extremely high yields. A few Sansai farmers are using this variety which won a recent national rice growing contest in the district (Saraphi) where Sansai is located. The winning value was 7.53 tons/ha. (or 1220 kg./*rai*).

Watabe describes the following ideal conditions for the winning entry in the national rice growing contest:

Map 9 *Late Rice Crop Choice at Sansai, 1967*

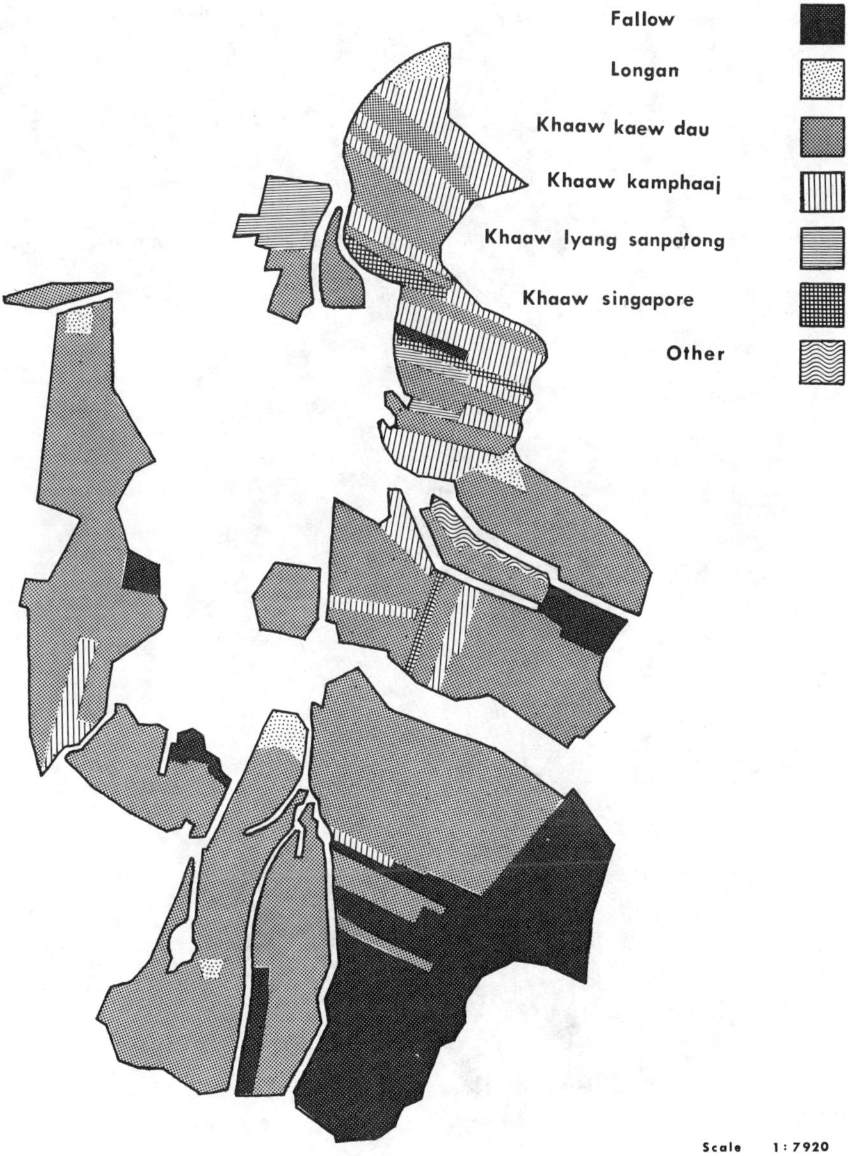

Map 10 *Late Rice Crop Choice at Sansai, 1968*

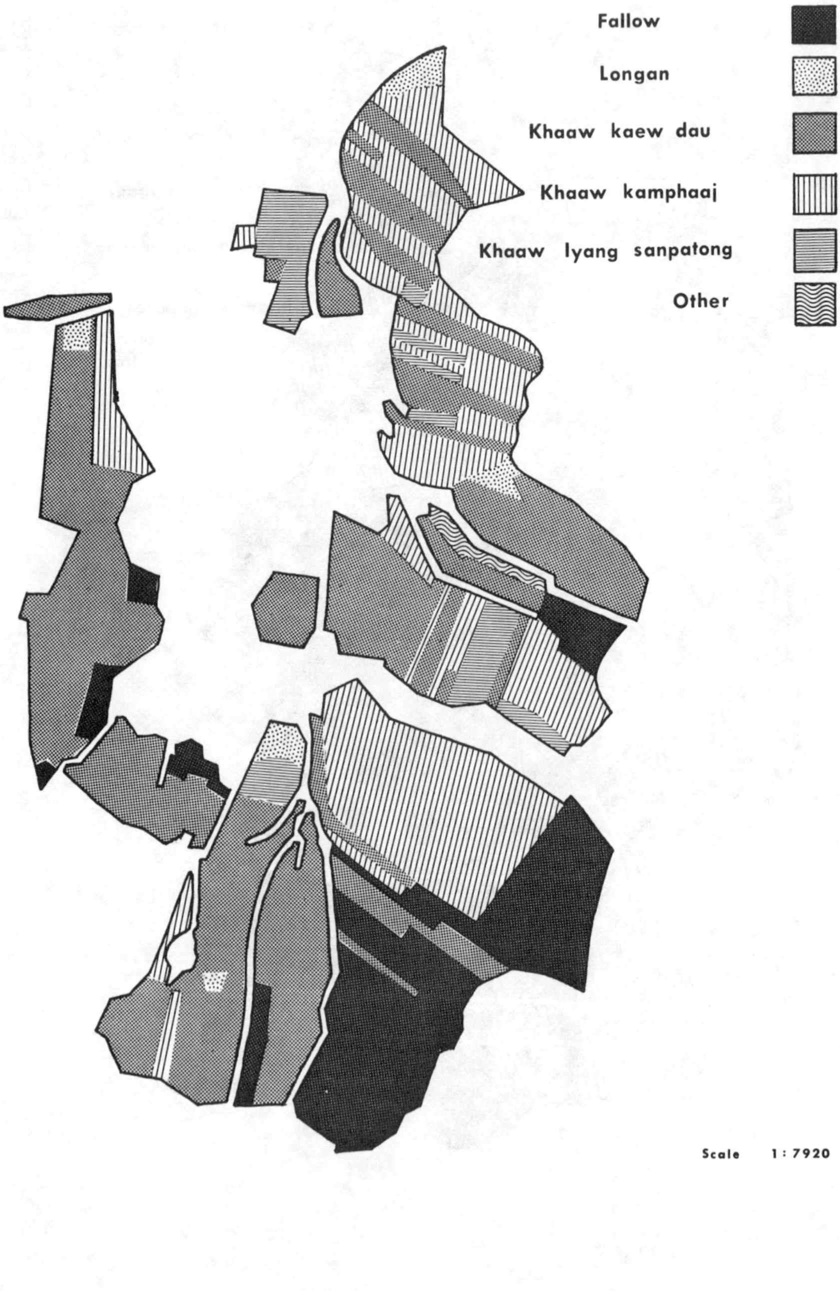

Map 11 *Late Rice Crop Choice at Sansai, 1969*

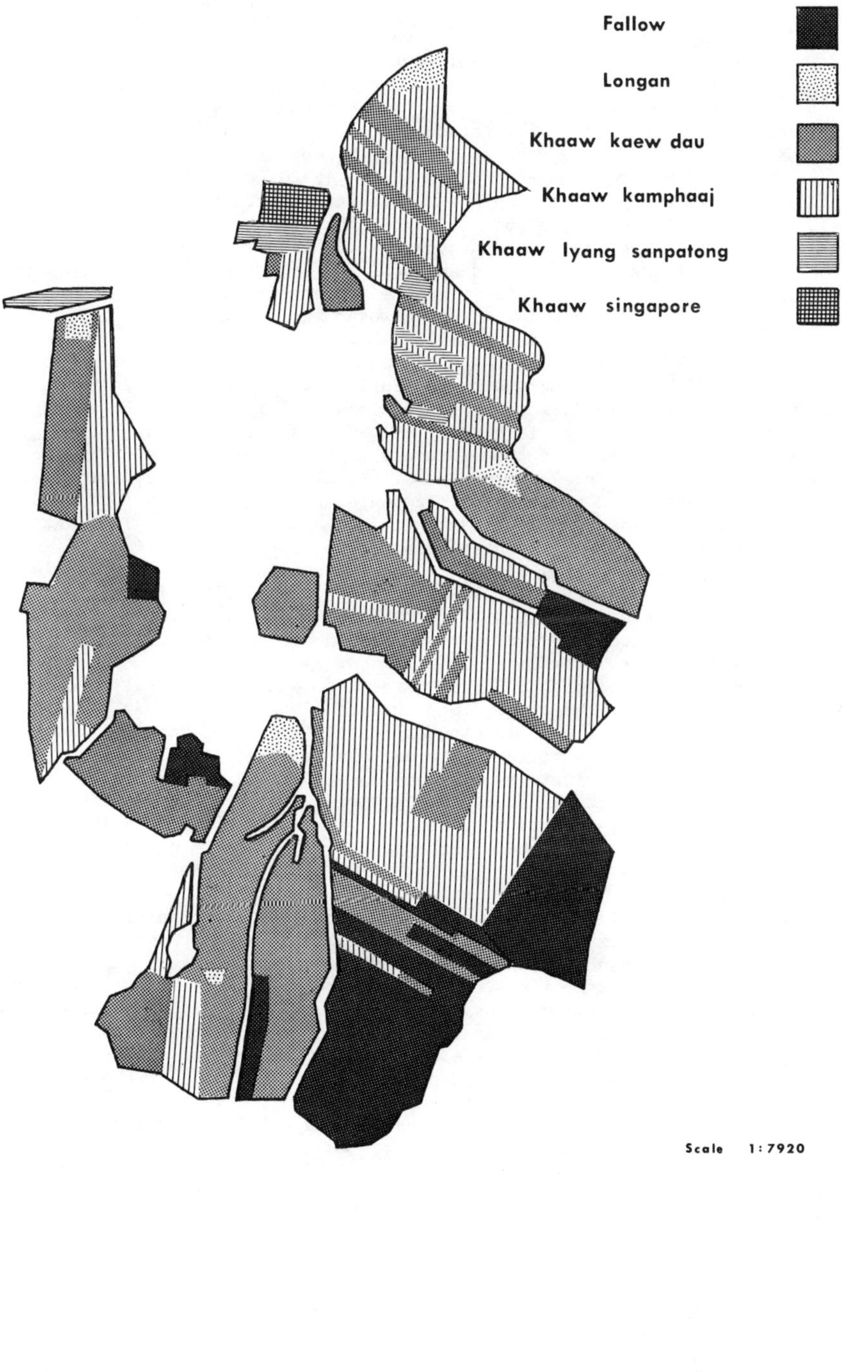

> This paddy field is located in a low basin on the left bank of Mae Nam Ping and is occasionally damaged by flood, although there are irrigation facilities. (In fact, in 1965, [the year after the contest] damage by flood brought about a greatly reduced yield.) Soil texture is clay loam and the surface soil is comparatively deep. Every year double cropping is carried on, and as a second crop soybeans, mungbeans or peanuts are cultivated. In 1964 it was soybeans, and after harvesting, the stalks were plowed under. In addition, 10 kg./*rai* ammo-phosphate (N:P = 16:20) was applied. From soil analysis, it was found that the amount of phosphate contained in the soil was exceptionally large. The variety planted was Niaw Sanpatong, sowed in the middle of July and transplanted in Mid-August. This planting took place about one month later than the average for Amphoe Saraphi. Following this, adequate water was supplied and there was no ill-effect (1967:117-118).

No Sansai household obtained comparable yields with *khaaw lyang San Pratong*. In 1969, yields were low--less than 200 kg./*rai* on the average. The fact that this variety is used at all in Sansai probably is because local farmers have heard of the extremely high possible yields. The fact that local fields are subject to regular flooding, and that no fertilizer is used no doubt limit potential yields in Sansai.

Very high yields are also possible with *khaaw kam phaaj*. In a national contest among Rice Experiment Stations, the Station at San Pratong (about ten kilometers from Sansai) placed second in Thailand with a stand of *khaaw kam phaaj* which yielded 5.22 tons/ha. or 845 kg./*rai*. This figure is also higher than any attained in Sansai for the 1969 late rice crop. It was attained through careful use of nitrogen and phosphate fertilizers, and presumably under ideal irrigation conditions.

The primary problem that Sansai farmers face in the production of the late rice crop is the threat of floods between late July and early October. In most years, at least one flood sufficient to cover the tops of transplanted seedlings in the village's lowest planted fields can be expected. 1967 is the only year within memory when no such flooding occurred. Sansai farmers have coped with the problems of raising two rice crops since the 19th century. In some years, more than one flood may occur. In 1971, letters from informants in the village reported that there was danger of flooding as early as mid-July. The waters subsided, but rose to flood stage again in August.

Flooding not only threatens much of the rice crop at Sansai, it also places heavy labor demands on the farmers. Sansai Sub-district and the sub-districts to the north and south are protected from direct flooding by the Ping river by a large levee. In recent years, Sansai fields have been protected from direct flooding by the levee, but a substantial problem with seepage and accumulated rainwater remains. Each year, Sansai farmers express the hope that levees on the other side of the river, or far downstream, will break first. This event would take pressure off their side of the river and protect their fields from direct flooding. In fact, direct flooding is seldom a problem, but the water accumulated through rainfall and seepage is sufficient to cause substantial damage.

The levee protecting Sansai's fields in 1969 was threatened by high waters of the Ping river as early as mid-August. Sansai farmers began strengthening the levee as early as August 20. The river was already two to three meters above the level of the land behind the barrier. Some farmers with land behind the levee on the opposite side (Hang Dong District) of the river had already suffered severe flooding while harvesting the early rice crop. They had a large amount of damp, recently harvested paddy which they were attempting to thresh. The stalks were threshed by holding them against the inside of the harvest basket, and beating them with a small branch from a coconut tree. The grain was dried in the sun. Because of dampness, some rice already had been lost through sprouting.

The river levee is reinforced at places subject to strong, swirling currents by driving bamboo poles (about 2.5 meters long and five cm. in diameter) into the river bottom next to the bank. The vertical poles, pounded in with large wooden hammers, are held in place by horizontal poles lashed to them. Other pieces are then tied to the cross-pieces and staked into the bank.

If the levee leaks, men dig a vertical hole down until they find the horizontal hole which is causing the leak. The leak is stopped with sandbags. On August 20, one of Sansai's two headmen had sent letters to the District office and the Irrigation Department, requesting from each 800 gunny sacks, suitable for use as sandbags.

On August 20, 1969 a Chiang Mai radio station announced that the Ping river was 3.52 meters above flood stage. In 1966, one of the worst flood years on record, the maximum figure was 3.82 meters. Indications are that the floods of 1969 were worse than usual. For example, people with fish ponds were forced to construct bamboo fences to retain their fish.

At a meeting on August 21, 1969, it was decided that each affected village--Sansai and seven others--would be required to send five to twenty men during each twenty-four hour period to act as guards at the levee. The number of guards depended on the village's size. Each guard was required to bring a length of bamboo (about 2.5 meters) and a bundle of rice straw. If possible, they were also to bring a large, machete-type knife, a wooden hammer, or a basket for carrying dirt. Each farmer with *kaa kaan* responsibility to a piece of land (that is, each owner or renter who accepts responsibility for work on irrigation schemes) was expected to take his turn at guarding the levee. In cases of real emergency--if the levee shows clear signs of breaking--all farmers in these villages were expected to turn out.

Within a week, the Ping river had fallen, and the danger was past. The flood had damaged fields in Sansai, and five other villages within the Sub-district. Farmers whose fields had sustained damage were busy sowing new nurseries or seeking seedlings to fill in for individual drowned plants. New seedlings were obtained from three sources. First, farmers with particularly low fields either waited until after the flood to sow (*waan haj*) or were forced to re-sow after the flood waters receded. Second, farmers fortunate to have high as well as low fields, often practice *kaa cam* for the late crop. When they transplant their higher fields, they transplant at "double density" near the edges of the field. In case of flooding, these "extra" plants can be used as very tall seedlings to replace individual drowned plants. In years when no flood damage is experienced, extra plants are pulled up after the flood danger is clearly past. Third, two villages to the north of Sansai have the highest fields in the Sub-district. A number of farmers in both villages raise seedlings for sale, anticipating that a certain amount of flooding will occur every year. A thickly seeded stand of seedlings two meters square is sold for 10 *baht*. This is sufficient to transplant roughly 0.5 *rai*.

On September 12, 1969, caterpillars were first noticed eating rice plants in areas that had been affected by the flood. Farmers with fields to the south of the village went to the District office to obtain insecticide from the official representing the Agriculture Ministry. They were given a five gallon container of a liquid insecticide called Guthion. Guthion, manufactured in the United States, is one of the dangerous, highly poisonous organic-phosphate insecticides. The container had several warning notices in English, informing the user of possible dangers from fumes, exposure of the skin, etc. There were no warnings in Thai. The farmer in charge of the insecticide denied that the Agriculture Ministry official had provided any verbal warning. This writer first noticed the container of Guthion and its many warnings in English because a number of children were playing around it.

Labor Expenditures: The Early Crop

Rice is cultivated in Sansai with relatively small inputs of capital. The tools, few and inexpensive, are used for several years before wearing out. Some are used in a variety of tasks, both within and outside of rice cultivation. Below is listed a "typical" inventory of tools and equipment (owned by a household with an agricultural labor force of two, and working 3.0 *rai*): one water buffalo (1500 *baht*); one plow (77 *baht*); two hoes (44 *baht*); one harrow (25 *baht*); and one harvest knife (8 *baht*).

The most expensive item on the above list is the water buffalo. However, its purchase price seems less high when one considers that the animal may be used for plowing and harrowing for two or three crops each year, for a total period of 10 or 15 years. The animal may also be rented to other farmers. A female will produce offspring with cash value. Finally, when the animal is too old to work it may be sold to local butchers. The greatest expense, other than land, that Sansai farmers have in producing a rice crop is labor. Table 12 summarizes available data on labor expenditures for the early rice crop at Sansai. The data collected at Sansai agree remarkably well with labor expenditure figures published in other sources. Watabe reports that northern Thai farmers are able to plow 1.0 *rai* in a day, and harrow 1.5 *rai* in a day (1967:66). Assuming that a field is plowed once, and harrowed two or three times, the "labor expenditure" for plowing and harrowing a *rai* of land should be roughly 2.3 - 3.0 man-days (and buffalo-days) per *rai*. The mean figure for Sansai is 2.97 days per *rai*.

In Taiwan, 10 men harvest 1.2 acres of paddy in a day. This labor expenditure is equivalent to 3.3 man-days per *rai* (Gallin 1966:49). The mean expenditure on reaping at Sansai is 3.64 man-days per *rai*. Hand weeding in Taiwan is done at the rate of one-fourth acre per worker per day. Assuming two weedings for each crop, an acre of wet rice would require eight man-days of weeding (Mendiola 1949:56). This labor expenditure is equivalent to 3.2 man-days per *rai*. The mean expenditure on weeding at Sansai is 4.53 man-days per *rai*.

In Hong Kong, one man can irrigate, examine and repair the bunds of 6-10 acres of wet rice in a day. Thus, a single acre would require about one-eighth of a day for each inspection. Assuming that 20 inspections are necessary each season, the labor requirement would be 2.5 days per acre or 1.0 day per *rai* (Herklots 1950:323). The mean labor expenditure on irrigation and bund repair in Sansai is 1.04 man-days per *rai*.

Table 12 *Labor Expenditures on the Early and Late Rice Crops at Sansai, 1969-70*

Operation	Early Crop		Late Crop	
	Mean Expenditure (Man-days/*rai*)	(SD)	Mean Expenditure (Man-days/*rai*)	(SD)
Preparing nursery bed	0.53	(0.35)	0.43	(0.29)
Watering nursery	0.72	(0.60)	0.62	(0.62)
Plowing/harrowing	3.14	(1.03)	3.28	(1.72)
"Cleaning" the field	1.29	(1.29)	2.03	(1.46)*
Pulling, cutting, and carrying the seedlings	1.86	(1.12)	1.50	(0.91)
Transplanting	2.47	(1.01)	2.51	(1.47)
Weeding	4.85	(2.38)	3.39	(2.60)*
Irrigation	1.41	(1.16)	2.34	(1.95)*
Reaping	3.91	(1.62)	3.27	(1.68)**
Threshing	1.96	(1.70)	1.21	(0.76)*
Carrying paddy to storage	2.10	(1.11)	1.40	(0.91)*
Various minor tasks	0.14	(0.29)	0.34	(0.73)
Total labor Expended	24.38	-	22.32	-

*Difference between crops significant at .01 level.

**Difference between crops significant at .05 level.

Sansai households that raise the early crop apply an average of 24.4 days to each *rai* between preparation of the nursery bed and carrying the threshed rice to storage. Equivalent figures for other wet rice raising areas in Asia are listed in Table 13. There is reason to assume that the figures calculated for Sansai are generally sound. Labor expenditures are somewhat higher than those for Ban Ping (in Chiang Rai Province, northern Thailand), for Nakhorn Pathom Province (in central Thailand), and Cochin China (south Vietnam), all of which are characterized by larger farms than those at Sansai. Buck's figures gathered in northern and central China are somewhat lower than those gathered in Sansai. Dumont's figure for northern Vietnam is little more than a crude extimate, backed by no survey data. It probably is unrealistically high. Gourou's figure for the Tonkin delta is based on a survey of households much smaller than that carried out in Sansai. Based on the data in Table 13 the information gathered on labor expenditures in Sansai appears accurate.

Decision Rules for the Early Rice Crop

A survey carried out after the early rice crop of 1970 was described earlier. Through quantitative analysis, reference to field/crop maps, and discussions with Sansai farmers, the following five major decision "rules" (and sub-rules) have been explicated for the early rice crop.

First, an average of 22-25 man-days of labor must be applied in the production of a single *rai* of early rice. (See Table 12.) This work includes eleven "major" and several minor tasks. Standard deviation values for each task indicate substantial variation between households in amount of labor devoted to each task. Yet assessment of total labor inputs under a variety of conditions suggests that inputs of 20-25 man-days per *rai* encompass the "ordinary" range of labor inputs. (See Table 14.)

Second, because they are relatively high, certain fields within the boundaries of Sansai are seldom or never planted in the early rice crop. The gravity operated irrigation system is incapable of providing these fields with adequate water in the early spring when early rice is planted. (See Map 5 and Chart 12.)

Third, the 1970 mean yield figures were related to: landholding status of the farming household; choice of rice variety; timing of agricultural operations, particularly initiation of transplanting; and total labor inputs, as well as labor inputs into the specific task of weeding. (See Table 15.) In 1970, renting households achieved *lower* yields for the early crop than did owning households. This difference probably is related to the fact that the early, nonglutinous rice crop is raised largely for sale. Households owning land which traditionally produces poorly during the early rice season may decide that, as a cash crop, early rice is not worth their labor investment, and either allow the land to remain fallow or rent it out on a half-share basis. Poorer, renting households cannot afford to make this choice.

In 1970, households planting *khaaw kaseet* received higher yields than those planting *khaaw lyang nauj*. *Khaaw kaseet* is a genetic cross between the "miracle" rice IR-8 and *khaaw lyang nauj*. Sansai farmers are convinced of its superiority, even without application of fertilizer. Yield figures for 1970 suggest that their confidence in varietal superiority is well-founded. In 1970, households that transplanted before April 10 received higher yields than those which planted later. This difference is probably explained by the fact that the latter households had the misfortune to transplant during a drought period which extended from April 9 to April 22, and which saw

Table 13 *Labor Expenditures on Wet Rice Cultivation in Various Asian Areas*

Area	Labor Expended (Days/*rai*	Labor Expended (Days/acre)	Labor Expended (Days/hectare)	Source
Sansai early rice	24.4	61.0	152.5	
Sansai late rice	22.3	55.8	139.4	
Ban Ping, northern Thailand	15.8	39.6	99	(Moerman 1968: 206)
Nakhorn Pathom, central Thailand	16.2	40.4	101	(Thailand 1961: Table 45)
Cochin China	13.6	34.0	85	(Gourou 1966: 360)
Tonkin	32	80	200	(Gourou 1966: 297)
Northern Vietnam	36	90	225	(Dumont 1965: 138)
Northern, eastern, and central China (non-glutinous	18.7	46.8	117	(Buck 1930: 227)
Northern, eastern, and central China (glutinous)	19.5	48.8	122	(Buck 1930: 227)

maximum daily temperatures varying between 39° and 44°C. Extreme weather of this type is undoubtedly hard on newly transplanted seedlings. Mean yields in 1970 are clearly related to total labor inputs. Households which devoted more labor to the crop ordinarily received higher yields. (See Tables 14 and 16.) A similar relationship exists between 1970 yields and labor inputs into the specific task of weeding. (See Tables 16 and 17.)

Other variables that were explored, but had no significant relationship to early rice yields in 1970 included: Socioeconomic Group; household labor force size; "labor intensity"; planting date; harvest date; and altitude of land planted.

Fourth, labor inputs into plowing/harrowing and weeding are significantly related to "labor intensity." (See Table 17.) That is, on the average, households with more workers per *rai* devote more time to these operations.

Fifth, expenditures on hired labor are significantly related to landholding status in 1970. (See Table 18.) For example, owning households hired significantly more man-days of labor than renting households.

In Chart 12 the five major decision "rules" just presented are recorded in graphic form. A number of important points are illustrated. First, rice yields for the early crop at Sansai are dependent on five major variables: altitude of land worked; landholding status of farming household; choice of variety; planting date; and labor inputs.

Table 14

Labor Expenditures Under Various Conditions for the Early Rice Crop, 1970

Conditions	Mean Expenditure (Total Man-days/*rai*)
Rice raised by:	
Socioeconomic Group I (N = 21)	23.59
Socioeconomic Group II (N = 20)	25.27
Socioeconomic Group III (N = 15)	24.30
Rice raised on:	
owned land (N = 20)	25.23
rented land (N = 32)	24.80
Rice raised with:	
1-2 household workers (N = 26)	22.97
3-4 household workers (N = 25)	25.82
5 or more household workers (N = 5)	25.92
Rice raised with:	
more than 1 household worker/*rai* (N = 20)	28.21
2/3-1 household worker/*rai* (N = 26)	22.75
fewer than 2/3 household worker/*rai* (N = 10)	20.87
Fields planted in:	
khaaw lyang nauj (N = 49)	24.33
khaaw kaseet (N = 7)	24.60
Rice planted:	
Feb. 2-Feb. 11 ("early") (N = 11)	25.48
Feb. 12-Feb. 27 ("medium") (N = 30)	23.16
Feb. 28-Mar. 25 ("late") (N = 12)	26.70
Rice transplanted:	
Mar. 3-Mar. 20 ("early") (N = 11)	24.38
Mar. 21-Apr. 8 ("medium") (N = 34)	25.14
Apr. 10-Apr. 25 ("late") (N = 11)	22.00
Rice raised on:	
"low" land (N = 7)	25.57
"medium" land (N = 30)	25.02
Fields producing:	
15-39 buckets/*rai* (N = 12)	21.71
40-59 buckets/*rai* (N = 28)	23.19
60-90 buckets/*rai* (N = 16)	28.48

Table 15 *Yield Figures Under Various Conditions for the Early Rice Crop, 1970*

Conditions	Mean Yield (buckets/*rai*)	(SD)
Rice raised by:		
Socioeconomic Group I (N = 21)	53.3	(14.3)
Socioeconomic Group II (N = 19)	40.6	(14.6)
Socioeconomic Group III (N = 15)	46.2	(13.3)
Rice raised on:		
owned land (N = 20)	54.8	(12.2)
rented land (N = 32)	47.8	(15.4)*
Rice raised with:		
1-2 household workers (N = 26)	46.5	(13.6)
3-4 household workers (N = 25)	51.3	(15.6)
5 or more household workers (N = 5)	58.0	(4.4)
Rice raised with:		
more than 1 household worker/*rai* (N = 19)	50.2	(13.7)
2/3-1 household worker/*rai* (N = 28)	47.8	(14.4)
fewer than 2/3 worker/*rai* (N = 9)	54.4	(14.8)
Fields planted in:		
khaaw lyang nauj (N = 49)	48.8	(14.8)
khaaw kaseet (N = 6)	57.5	(10.4)**
Rice planted:		
Feb. 2-Feb. 11 ("early") (N = 11)	48.0	(14.2)
Feb. 12-Feb. 27 ("medium") (N = 30)	49.4	(12.4)
Feb. 28-Mar. 25 ("late") (N = 12)	49.2	(17.0)
Rice transplanted:		
Mar. 3-Mar. 20 ("early") (N = 11)	51.0	(14.7)
Mar. 21-Apr. 8 ("medium") (N = 34)	51.4	(14.5)*
Apr. 10-Apr. 25 ("late") (N = 11)	43.1	(12.0)
Rice raised on:		
"low" land (N = 7)	56.6	(12.8)
"medium" land (N = 33)	50.4	(14.0)

*Significant at .05 level.
**Significant at .10 level.

Ordinarily, the farming household has little control over the first two variables. Land has become so expensive in Sansai that it is seldom purchased. Ownership of land is largely a chance event, based on inheritance and birth into the right family. The same is true of high, medium, or low land.

Table 16 *The Relationship Between Yield and Labor Expenditures on Various Tasks During the Early Rice Crop in Sansai, 1970*

Task	Low Yield 15–39 Buckets per *Rai* (N = 12)	(SD)	Medium Yield 40–59 Buckets per *Rai* (N = 28)	(SD)	High Yield 60–90 Buckets per *Rai* (N = 16)	(SD)
Preparing nursery bed	0.59	(0.50)	0.51	(0.29)	0.50	(0.32)
Watering nursery	0.83	(0.57)	0.53	(0.36)	0.98	(0.81)**
Plowing/harrowing	2.90	(1.36)	3.09	(0.73)	3.38	(1.14)
"Cleaning" the field	0.78	(0.84)	1.36	(1.28)	1.54	(1.43)
Pulling, cutting, and carrying the seedlings	1.98	(1.37)	1.66	(0.67)	2.11	(1.43)
Transplanting	2.45	(0.87)	2.37	(0.94)	2.66	(1.19)
Weeding	3.55	(2.08)	4.63	(2.15)	6.22	(2.28)*
Irrigation	1.30	(1.27)	1.18	(1.06)	1.89	(1.10)
Reaping	3.69	(1.30)	3.54	(1.48)	4.72	(1.76)
Threshing	1.67	(0.92)	2.09	(2.25)	1.98	(0.75)
Carrying paddy to storage	1.63	(0.74)	2.07	(2.21)	2.49	(1.01)
Various minor tasks	0.14		0.16		0.11	
Total labor expended	21.51		23.19		28.58	
Mean Yield	30.6 buckets		46.8 buckets		66.8 buckets	

*Significant at .01 level.
**Significant at .05 level.

95

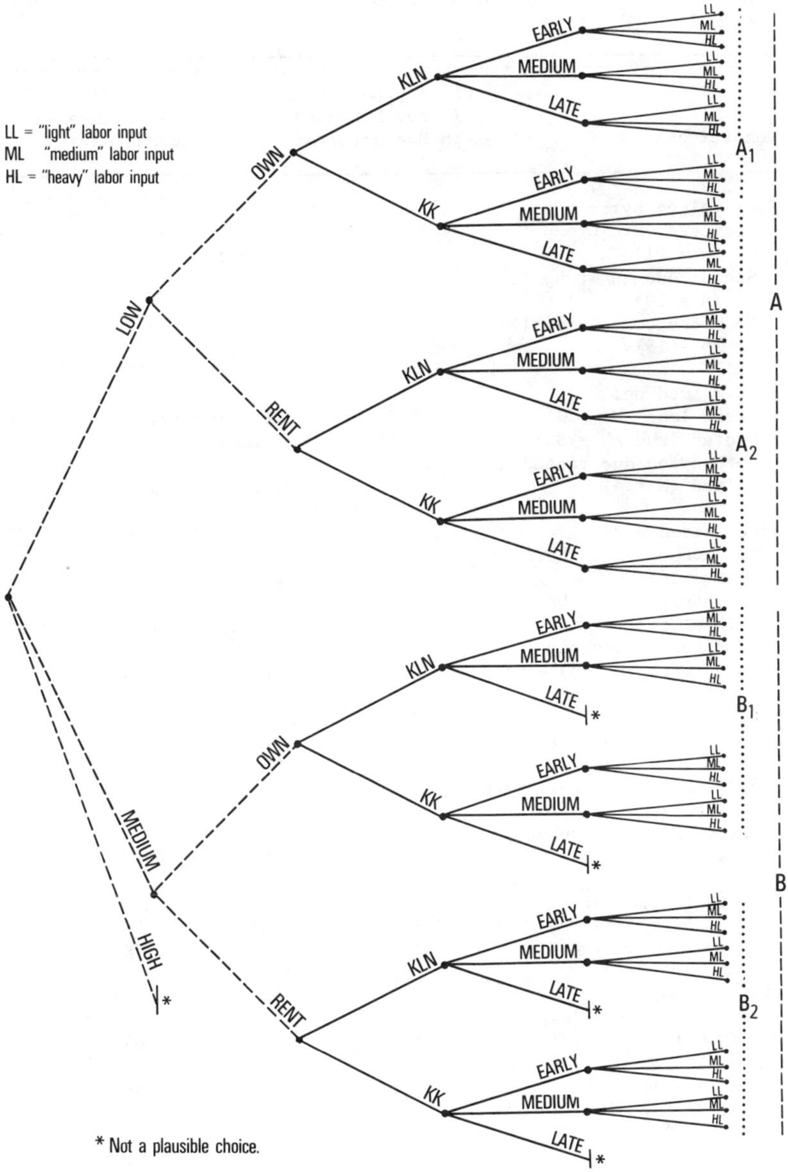

Chart 12 *Decision Tree for Early Rice*

Table 17 *Labor Expended on Plowing/Harrowing and Weeding Under Various Conditions for the Early Rice Crop in Sansai, 1970*

Conditions	Plowing/Harrowing (Labor Expended in Man-days/*rai*	(SD)	Weeding (Labor Expended) in Man-days/*rai*	(SD)
Rice raised by:				
Socioeconomic Group I (N = 21)	3.09	(1.30)	5.41	(2.72)
Socioeconomic Group II (N = 20)	3.06	(0.84)	4.44	(2.50)
Socioeconomic Group III (N = 15)	3.28	(0.78)	4.61	(1.40)
Rice raised on:				
owned land (N = 20)	3.09	(1.01)	4.89	(2.09)
rented land (N = 32)	3.21	(1.08)	4.96	(2.54)
both owned and rented land (N = 4)	2.80	(0.54)	3.83	(2.14)
Rice raised with:				
1-2 household workers (N = 26)	3.00	(0.96)	4.03	(1.67)
3-4 household workers (N = 25)	3.26	(1.08)	5.60	(2.66)
5 or more household workers (N = 5)	3.24	(1.05)	5.40	(2.58)
Rice raised with:				
more than 1 household worker/*rai* (N = 20)	3.65	(1.11)	6.26	(2.60)
2/3-1 household worker/*rai* (N = 26)	2.84	(0.99)**	4.00	(1.55)*
fewer than 2/3 household worker/*rai* (N = 10)	2.87	(0.25)	4.25	(2.38)
Fields planted in:				
khaaw lyang nauj (N = 49)	3.15	(0.98)	4.86	(2.42)
khaaw kaseet (N = 7)	3.01	(1.35)	4.79	(2.00)
Rice planted:				
Feb. 2-Feb. 11 ("early") (N = 11)	3.41	(0.64)	5.09	(2.28)
Feb. 12-Feb. 27 ("medium") (N = 30)	2.82	(1.08)	4.58	(2.12)
Feb. 28-Mar. 25 ("late") (N = 12)	3.60	(1.04)	5.35	(2.90)

Table 17 (Continued)

Labor Expended on Plowing/Harrowing and Weeding Under Various Conditions for the Early Rice Crop in Sansai, 1970

Conditions	Plowing/Harrowing (Labor Expended in Man-days/*rai*	(SD)	Weeding (Labor Expended) in Man-days/*rai*	(SD)
Rice transplanted:				
Mar. 3-Mar.20 ("early") (N = 11)	3.31	(0.67)	5.25	(2.59)
Mar. 21-Spr. 8 ("medium) (N = 34)	3.08	(0.99)	4.79	(2.37)
Apr. 10-Apr. 25 ("late") (N = 11)	3.14	(1.36)	4.65	(2.10)
Rice raised on:				
"low" land (N = 7)	3.12	(1.24)	5.49	(2.18)
"medium" land (N = 30)	3.01	(0.50)	3.76	(2.25)
Fields producing:				
15-39 buckets/*rai* (N = 12)	2.90	(1.30)	3.55	(2.08)
40-59 buckets/*rai* (N = 28)	3.09	(0.84)	4.63	(2.15)*
60-90 buckets/*rai* (N = 16)	3.28	(0.78)	6.22	(2.28)

*Significant at .01 level.
**Significant at .05 level.

The heir must receive whatever land he receives, and the renter accepts land wherever he finds it. The owner of high land must accept the fact that he is unable to produce early rice, while the renter must accept the fact that he receives only one-half of what he produces.

The owner *does* have control over rice variety, timing, and labor inputs. Combining these elements in various ways provides each household with a wide range of potential yields. Chart 12 suggests the existence of twelve or eighteen alternate payoffs for each household. In fact, this Chart is too simple. "Choice of variety" can reasonably be divided into discrete categories. But "timing" and "labor inputs" are divided into discrete categories only for convenience. Each of these variables comprises an infinite gradation of choices. So the list of possible pay offs is infinite--albeit with definite upper and lower limits.

If one considered additional variables--too subtle for consideration in this survey--such as soil type, water regime, and previous crops, it would become evident that each household is faced with a *unique* decision-making situation. However, it is also evident that one can group individual decision matrices into categories which represent "micro-environmental" variation in significant ways. Decisions against Nature belonging to Category A (see Chart 12) are effected on Sansai's lowest fields. Households using these fields raise early rice in the full knowledge that no late rice

Table 18 *Use of Hired Labor Under Various Conditions for the Early Rice Crop, 1970*

Condition	Mean Expenditure (Man-days/*rai*)	(SD)
Rice raised by:		
Socioeconomic Group I (N = 21)	3.76	(3.29)
Socioeconomic Group II (N = 20)	5.05	(4.58)
Socioeconomic Group III (N = 15)	2.02	(1.96)
Rice planted:		
Feb. 2-Feb. 11 ("early") (N = 11)	3.97	(4.96)
Feb. 12-Feb. 27 ("medium") (N = 30)	4.00	(3.38)
Feb. 28-Mar. 25 ("late") (N = 12)	2.58	(2.95)
Rice transplanted:		
Mar. 3-Mar. 18 ("early") (N = 10)	3.47	(4.88)
Mar. 20-Apr. 10 ("medium") (N = 35)	4.45	(3.35)
Apr. 11-Apr. 25 ("late") (N = 11)	1.61	(2.27)
Rice raised with:		
more than 1 household worker/*rai* (N = 20)	2.62	(2.53)
2/3-1 household worker/*rai* (N = 26)	4.53	(3.56)
fewer than 2/3 household worker/*rai* (N = 10)	3.42	(5.27)
Rice raised on:		
owned land (N = 20)	5.00	(4.02)
rented land (N = 32)	3.05	(3.31)*
Fields producing:		
200-300 kg./*rai* (N = 11)	3.83	(3.29)
400-590 kg./*rai* (N = 28)	3.55	(4.00)
600-1050 kg/*rai* (N = 17)	3.86	(3.42)
Rice variety:		
khaaw lyang nauj (N = 49)	4.01	(3.84)
khaaw kaseet (N = 7)	1.50	(0.82)
Rice raised with:		
1-2 household workers (N = 26)	4.52	(4.13)
3-4 household workers (N = 25)	3.21	(3.35)
5 or more household workers (N = 5)	1.88	(0.87)

*Significant at .05 level.

crop will follow. Decisions against Nature in Category B are effected on fields of medium height. Households using these fields are subject to at least one major constraint which does not apply to Category A. If the early rice crop is planted too late, any chance of planting a late rice crop will be eliminated. Owners of high land are not allowed to choose the "early rice" strategy.

Graphic representation of possible strategies in early rice cultivation reemphasizes one already apparent point, and emphasizes two less obvious ones. The obvious point is that Sansai renters are working toward payoffs radically altered from those available to owner-operators. Their payoffs are only one-half of what they produce.

One less obvious point is that within a single, small village, minor variations in environment--centering on altitude and water supply--have a major effect on agricultural decision-making. In effect, households working high, medium, and low land are faced with different decision frameworks. The second point is that the early rice crop does not, in itself, constitute a complete decision framework. Its strategies cannot be completely defined without reference to the succeeding crop. Only by reference to the following, late rice crop does one realize the unreasonableness of planting early rice late on land of medium altitude.

Decision Rules for the Late Rice Crop

A survey carried out after the 1969/70 late rice crop has been described. Through quantitative analysis, reference to field/crop maps, and discussions with Sansai farmers, the following five decision "rules" (and subrules) have been explicated for the late rice crop.

First, an average of 22-25 man-days of labor must be applied in the production of a single *rai* of late rice. (See Table 12.) Total labor inputs for the early and late crops are nearly the same. However, significant differences *do* exist for several individual operations. The early rice crop requires significantly greater inputs into reaping, threshing, and carrying paddy to storage. Almost certainly, this reflects the higher average yields that are achieved with the early crop. More grain requires more harvest labor. The late crop requires significantly more labor for "cleaning" the fields, while the early crop requires significantly more labor for weeding. However, when labor expenditures on these similar operations are combined, the resulting values--6.14 man-days per *rai* for the early crop and 5.42 man-days per *rai* for the late crop--are quite similar.

Second, because they are relatively low, certain fields within Sansai are seldom or never planted with late rice. Because drainage is inadequate, these fields are covered by slowly rising (later, slowly falling) flood waters during the period from late July to early October. The greatest depth reached on the lowest field in Sansai is about 1.5 meters. (See Map 5 and Chart 13.)

Third, for the late rice crop in Sansai, 1969/70 mean yield figures were significantly related to: land-holding status of the farming household; choice of rice variety; and timing of agricultural operations, including planting and transplanting. (See Table 20.) In 1969/70, renting households received significantly *higher* yields than owning households. This difference (directly opposed to that described above for the early crop) is probably related to the fact that glutinous rice raised during the late crop is the primary source of local food. Land owning households may well decide that since adequate production will allow them to avoid all cash expenditures on rice, cultivation of even relatively poor fields is worth their labor.

In 1969/70, Sansai households planting *khaaw kaew dau* received significantly lower yields than those planting *khaaw kam phaaj*. These are the two varieties most often planted in Sansai. *Khaaw lyang San Pratong* is planted occasionally in Sansai, and in 1969/70 it provided average yields superior to *khaaw kaew dau* and inferior to *khaaw kam phaaj*. A number of households planted a combination of varieties--ordinarily *khaaw kam phaaj* at the lower end of the field and *khaaw kaew dau* at the upper end. This strategy provided average yields virtually identical to those achieved by planting *khaaw kam phaaj* only. In 1969/70, Sansai households which transplanted between August 8 and August 31 received significantly higher yields than those which transplanted earlier or later. However, Table 20 does not indicate that a large number of households were forced to replant after their newly transplanted seedlings drowned in August floods. Transplanting between August 8 and August 31 was a good strategy for those households using land high enough to escape flood damage. The problem is, of course, that "high enough" varies from year to year. Only households controlling relatively high fields are able to transplant successfully in August in virtually all years. Other variables that were explored, but had no significant relationship to late rice yields in 1969/70 included: Socioeconomic Group; household labor force size, "labor intensity;" and altitude of land cultivated. Although the total labor expenditures were significantly related to yield for the early rice crop, this was not true for the late rice crop. (See Table 21.)

Fourth, labor inputs into weeding are significantly related to labor intensity and timing. (See Table 22.) That is, households with more workers per *rai* devote more time to keeping their fields weed free. Late starting households devote relatively little time to weeding. Perhaps the floods of late summer drown weeds, as well as rice seedlings.

Fifth, expenditures on hired labor are significantly related to landholding status. (See Table 23.) That is, owning households hire significantly more man-days of labor than renting households.

Chart 13 presents these five rules in a decision tree form. Late rice yields in Sansai are dependent on four major variables: altitude of land worked; landholding status of farming household; choice of variety; and planting date. The farming household has little control over the first two variables but they do over the last two. In selection of a rice variety, or even a combination of varieties, the farming household faces a discrete choice. But when one considers that varying amounts of land can be planted in each variety, varietal choice becomes a matter of infinite gradations. The same is true of the variable of timing. One is faced with an infinite list of possible payoffs, again with definite upper and lower limits.

For purposes of discussion, an infinite number of possible strategies can be usefully divided into finite categories. As with the early crop, distinctions between high, low, and medium land are important. Households controlling low land are excluded from the late crop. Late summer floods make planting impossible. Medium lands are usable for both rice crops. But this fact, in itself, imposes strategic restrictions. Only if the early crop is planted relatively early, and the late crop is planted relatively late, is it possible to raise both. High lands are planted without significant

Table 19 *Labor Expenditures Under Various Conditions for the Late Rice Crop, 1969*

Conditions	Mean Expenditure (Total Man-days/*rai*)
Rice raised by:	
Socioeconomic Group I (N = 35)	22.68
Socioeconomic Group II (N = 34)	19.21
Socioeconomic Group III (N = 15)	27.40
Rice raised on:	
owned land (N = 39)	22.22
rented land (N = 44)	22.52
both owned and rented land (N = 10)	21.93
Rice raised with:	
1-2 household workers (N = 46)	23.15
3-4 household workers (N = 40)	21.81
5 or more household workers (N = 7)	25.64
Rice raised with:	
more than 1 household worker/*rai* (N = 19)	28.07
2/3-1 household worker/*rai* (N = 37)	22.67
fewer than 2/3 household worker/*rai* (N = 37)	19.05
Fields planted in:	
khaaw kaew dau (N = 36)	24.44
khaaw kamphaaj (N = 29)	20.48
khaaw singkapau (N = 5)	22.14
more than one variety (N = 23)	21.34
Rice planted:	
May 20-June 31 ("early") (N = 29)	24.87
July 1-Aug. 14 ("medium") (N = 23)	24.44
Sept. 1-Oct. 10 ("late") (N = 44)	19.65
Rice transplanted:	
June 31-Aug. 5 ("early") (N = 31)	23.69
Aug. 8-Aug. 31 ("medium") (N = 18)	26.52
Oct. 3-Nov. 26 ("late") (N = 44)	19.65
Rice raised on:	
"medium" land (N = 51)	22.93
"high" land (N = 12)	21.97
Fields producing:	
2-19 buckets/*rai* (N = 25)	20.55
20-34 buckets/*rai* (N = 46)	21.98
35-80 buckets/*rai* (N = 22)	23.88

Table 20 *Yield Figures Under Various Conditions for the Late Rice Crop, 1969*

Conditions	Mean Yield (buckets/*rai*)	(SD)
Rice raised by:		
Socioeconomic Group I (N = 35)	29.3	(13.6)
Socioeconomic Group II (N = 33)	26.8	(11.9)
Socioeconomic Group III (N = 18)	28.8	(20.8)
Rice raised on:		
owned land (N = 39)	24.9	(10.8)
rented land (N = 44)	30.0	(16.7)**
Rice raised with:		
1-2 household workers (N = 46)	28.5	(15.7)
3-4 household workers (N = 40)	26.0	(13.6)
5 or more household workers (N = 7)	33.6	(10.6)
Rice raised with:		
more than 1 household worker/*rai* (N = 18)	29.6	(12.9)
2/3-1 household worker/*rai* (N = 42)	27.8	(12.6)
fewer than 2/3 household worker/*rai* (N = 33)	26.9	(17.5)
Fields planted in:		
khaaw kaew dau (N = 36)	21.5	(10.1)
khaaw kamphaaj (N = 29)	32.2	(11.8)*
khaaw lyang san pratong (N = 5)	27.0	(17.5)
more than one variety (N = 23)	32.3	(18.8)
Rice planted:		
May20-June 31 ("early") (N = 29)	27.5	(11.2)
July 1-Aug. 14 ("medium") (N = 20)	34.5	(15.3)**
Sept. 1-Oct. 10 ("late") (N =44)	25.2	(15.5)
Rice transplanted:		
June 31-Aug. 5 ("early") (N = 31)	25.3	(8.4)
Aug. 8-Aug. 31 ("medium") (N = 18)	38.4	(15.8)*
Oct. 3-Nov. 26 ("late") (N = 44)	25.2	(15.5)
Rice raised on:		
"medium" land (N = 51)	26.2	(13.9)
"high" land (N = 12)	30.1	(10.0)

*Significant at .01 level.

**Significant at .10 level.

Table 21 *Relationship Between Yield and Labor Expenditures on Various Tasks During the Late Rice Crop in Sansai, 1969*

	Low Yield (N = 25)	(SD)	Medium Yield (N = 46)	(SD)	High Yield (N = 22)	(SD)
Preparing nursery bed	0.42	(0.27)	0.46	(0.31)	0.39	(0.27)
Watering nursery	0.55	(0.42)	0.69	(0.74)	0.56	(0.50)
Plowing/harrowing	3.37	(2.13)	3.03	(1.44)	3.73	(1.65)
"Cleaning" the field	2.08	(1.23)	2.03	(1.48)	1.96	(1.66)
Pulling, cutting, and carrying the seedlings	1.38	(0.82)	1.64	(1.04)	1.37	(0.63)
Transplanting	2.36	(1.01)	2.60	(1.68)	2.50	(1.39)
Weeding	3.61	(3.21)	3.12	(2.29)	3.69	(2.34)
Irrigation	1.74	(1.11)	2.46	(2.24)	2.75	(1.88)
Reaping	3.42	(2.04)	3.01	(1.51)	3.64	(1.45)
Threshing	1.04	(0.99)	1.27	(0.74)	1.25	(0.41)
Carrying paddy to storage	1.30	(1.06)	1.29	(0.78)	1.74	(0.91)
Various minor tasks	0.27	-	0.39	-	0.30	-
Total labor expended	20.55	-	21.98	-	23.88	-
Mean Yield	12.3 buckets		26.0 buckets		48.2 buckets	

Table 22 *Labor Expended on Plowing/Harrowing and Weeding Under Various Conditions for the Late Rice Crop in Sansai, 1969*

Conditions	Plowing/Harrowing (Labor Expended in Man-days/*rai*)	(SD)	Weeding (Labor Expended in Man-days/*rai*)	(SD)
Rice raised by:				
Socioeconomic Group I	3.14	(1.49)	3.51	(1.96)
Socioeconomic Group II	2.94	(1.33)	2.58	(1.70)**
Socioeconomic Group III	3.79	(2.37)	4.57	(4.01)
Rice raised on:				
owned land (N = 39)	2.98	(1.59)	3.53	(2.48)
rented land (N = 44)	3.63	(1.86)	3.23	(2.75)
both owned and rented land (N = 10)	2.98	(1.21)	3.54	(2.28)
Rice raised with:				
1-2 household workers (N = 46)	3.40	(1.81)	3.30	(2.85)
3-4 household workers (N = 40)	3.20	(1.68)	3.17	(2.24)
5 or more household workers (N = 7)	3.01	(1.15)	5.19	(1.93)
Rice raised with:				
more than 1 household worker/*rai* (N = 19)	3.16	(0.98)	4.91	(2.99)
2/3-1 household worker/*rai* (N = 37)	3.62	(2.22)	3.41	(2.92)*
fewer than 2/3 household worker/*rai* (N = 37)	3.02	(1.36)	2.58	(1.40)
Fields planted in:				
khaaw kaew dau (N = 36)	3.41	(1.96)	3.88	(3.02)
khaaw kamphaaj (N = 29)	3.56	(1.65)	2.70	(1.57)
khaaw singkapau (N = 5)	2.46	(0.89)	3.16	(1.11)
more than one variety (N = 23)	2.91	(1.39)	3.54	(2.93)
Rice planted:				
May 20-June 31 ("early") (N = 29)	2.59	(0.77)	5.12	(3.18)
July 1-Aug. 14 ("medium") (N = 23)	3.31	(1.98)**	3.43	(2.31)*
Sept. 1-Oct. 10 ("late") (N = 44)	3.73	(1.89)	2.23	(1.34)

Table 22 (Continued) *Labor Expended on Plowing/Harrowing and Weeding Under Various Conditions for the Late Rice Crop in Sansai, 1969*

Conditions	Plowing/Harrowing (Labor Expended in Man-days/*rai*)	(SD)	Weeding (Labor Expended in Man-days/*rai*)	(SD)
Rice transplanted:				
June 31-Aug. 5 ("early") (N = 31)	2.67	(1.12)	4.33	(2.66)
Aug. 8-Aug. 31 ("medium") (N = 18)	3.24	(1.80)**	4.61	(3.46)*
Oct. 3-Nov. 26 ("late") (N = 44)	3.73	(1.89)	2.23	(1.34)
Rice planted on:				
"medium" land (N = 51)	3.36	(1.95)	3.18	(2.21)
"high" land (N = 12)	3.40	(0.91)	3.02	(2.47)
Fields producing:				
2-19 buckets/*rai* (N = 25)	3.37	(2.13)	3.61	(3.21)
20-34 buckets/*rai* (N = 46)	3.03	(1.44)	3.12	(2.29)
35-80 buckets/*rai* (N = 22)	3.73	(1.65)	3.69	(2.34)

*Significant at .01 level.

**Significant at .05 level.

Table 23 *Use of Hired Labor Under Various Conditions for the Late Rice Crop, 1969*

Conditions	Mean Expenditure (Man-days/*rai*)	(SD)
Rice raised by:		
Socioeconomic Group I	3.59	(3.88)
Socioeconomic Group II	2.38	(2.25)
Socioeconomic Group III	2.69	(4.60)
Rice planted:		
May 20-June 31 ("early") (N = 29)	3.71	(4.17)
July 1-Aug. 14 ("medium") (N = 25)	3.31	(4.98)
Sept. 1-Oct. 10 ("late") (N = 43)	2.38	(2.79)
Rice transplanted:		
June 31-Aug. 5 ("early") (N = 31)	3.26	(3.44)
Aug. 8-Aug. 31 ("medium") (N = 18)	2.38	(2.79)
Oct. 3-Nov. 26 ("late") (N = 44)	2.38	(2.79)
Rice raised with:		
more than one household worker/*rai* (N = 19)	2.28	(3.18)
2/3-1 household worker/*rai* (N = 42)	2.90	(3.96)
fewer than 2/3 household worker/*rai* (N = 33)	3.44	(3.88)
Rice raised on:		
owned land (N = 40)	4.06	(4.39)
rented land (N = 44)	2.10	(3.25)*
Fields producing:		
20-190 kg./*rai* (N = 24)	2.90	(4.66)
200-340 kg./*rai* (N = 47)	2.72	(3.66)
350-800 kg./*rai* (N = 22)	3.58	(2.97)
Fields planted in:		
khaaw kaew dau (N = 36)	3.46	(4.28)
khaaw kan phaaj (N = 29)	2.88	(3.85)
khaaw singkapau (N = 5)	3.02	(2.70)
more than one variety (N = 23)	2.31	(3.02)
Rice raised on:		
"medium" land (N = 51)	3.17	(2.21)
"high" land (N = 12)	3.02	(2.47)
Rice raised with:		
1-2 household workers (N = 46)	3.74	(4.69)
3-4 household workers (N = 40)	2.34	(2.59)
5 or more household workers (N = 7)	1.53	(1.35)

*Significant at .01 level.

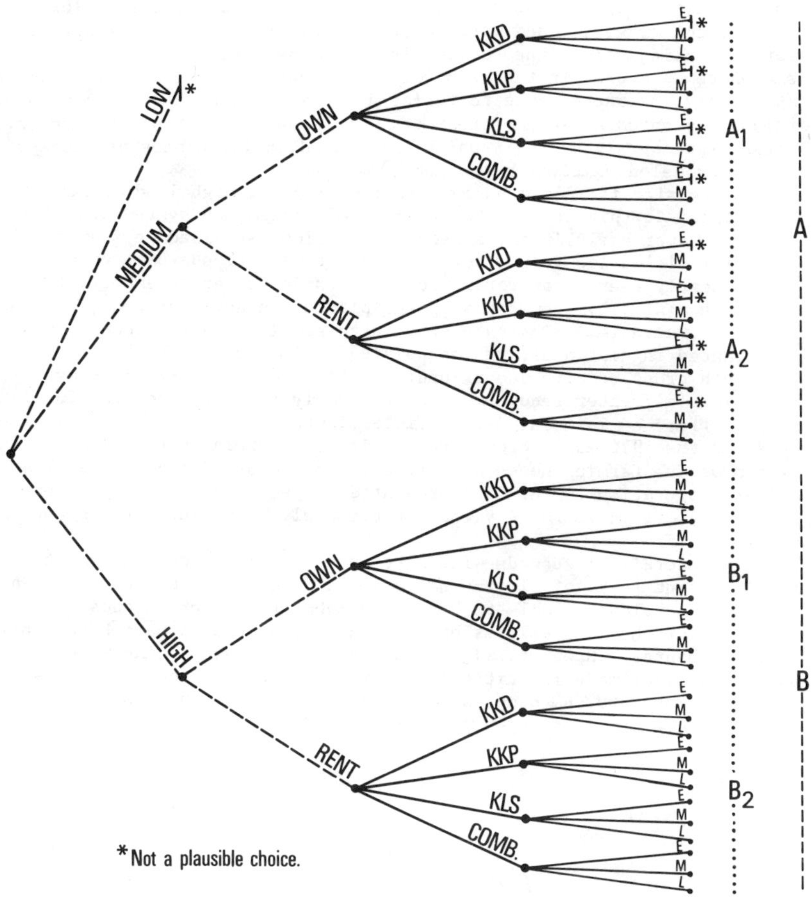

Chart 13 *Decision Tree for Late Rice*

restrictions as to time, because they lie fallow from March or April. Further divisions can be made on the basis of land holding. Renters face the fact that they keep only half of what they produce.

Crucial features to be emphasized are much the same as those indicated for the early rice crop. Owning and renting households face radically different decision-making frameworks. The same is true of farming households which work high, medium, and low lands. It is also the case that decision-making for the late crop is constrained by the requirements of the following crops. In particular, garlic and watermelons are planted in November or December. Later planting raises the possibility of inadequate water and extreme, hot season temperatures in the weeks before harvest. Since soybeans and peanuts apparently are better able to resist desiccation and extreme temperatures, they are planted as late as February. Households which plant late (e.g., after September 1) have virtually eliminated their chances of successful garlic or melon cultivation on the same land.

Early rice is raised on low and medium land; high land at Sansai is not planted in this crop. The gravity-fed irrigation system provides inadequate water early in the season. Late rice is raised on medium and high land; low fields lie fallow during this period. Floods have covered these fields nearly every year for as long as residents can remember. The early crop is essentially "commercial," and planted in non-glutinous, non-photosensitive varieties. The late crop is raised largely for local consumption, and planted in glutinous, photosensitive varieties.

Both crops require labor inputs on the order of 20-25 man-days per *rai*. The slightly greater requirement of the early crop is accounted for largely by difference in mean yield. Yield figures for both the 1969/70 late crop and the 1970 early crop were significantly related to: landholding status of the farming household; rice variety chosen; and timing of agricultural operations. The data presented suggest that heavy labor expenditures pay off in improved yield for the early crop, but not for the late crop.

In general, households with more workers per *rai* devote more time to the important tasks of plowing/harrowing and weeding. During both seasons, owning households are able to hire more labor than renting households.

Some of the complexities of rice farming at Sansai have been outlined in this Chapter. However, significant questions remain unanswered. First, how are choices made in critical situations? Second, how are choice procedures related to the process of adaptation? Tentative answers to these questions are provided in Chapters Seven and Eight.

Chapter Six

STRATEGIES IN GARDEN CROP CULTIVATION

This Chapter examines briefly the agronomic requirements of garlic, watermelons, soybeans, and peanuts as necessary background for understanding strategies used by Sansai farmers. Through maps, charts, tables, and statistical analysis, garden cropping strategies available to Sansai farmers are described.

During the research period, Sansai farmers had the opportunity to choose among four garden crops. Soybeans were introduced about 20 years ago. Garlic and peanuts have been important cash crops for several decades. Watermelons have been raised at Sansai for only three of four years. When such factors as labor requirement, fertilizer requirement, yield, and expected profit are considered, each crop has some advantages and disadvantages.

Garlic

At Sansai, garlic is planted and cultivated according to a distinct pattern. Garlic cloves are planted in flat beds, 2.0-2.5 meters wide, with ditches, about 25-30 cm. wide, between the beds. After treatment with fertilizer, the beds are covered with three to five centimeters of rice straw. The straw is thought to limit weed growth, and Sansai farmers never raise garlic without it. The plants push up through the straw; when brown they are harvested. Sansai farmers market their garlic immediately, in the form of *haum dip* (fresh garlic), or hold it for weeks or months, and sell it as *haum haeng* (dried garlic). Garlic sold fresh must be cut and washed on the day of harvest. Dried garlic is tied in small bundles and hung to dry in the rafters or under the raised floors of the house or rice storage bin.

Garlic cultivation can be divided into twelve major tasks: purchase of fertilizer; preparation of garlic seed; plowing; preparation of beds and ditches; planting garlic seed and preparation of straw; application of fertilizer and straw; irrigation; weeding (and application of additional fertilizer); harvest, carrying from field, preparation for sale, and selling; drying; bundling; and storage. Each of these tasks will be considered in turn.

Purchase of Fertilizer

Sansai farmers use *puj daeng* (a combination of nitrogen, phosphate, and potassium) or *puj khaaw* (urea) or a combination of the two. *Puj daeng* costs 95-105 *baht* per 50 kg. bag. A 50 kg. bag of *puj khaaw* costs 70 *baht*. Fertilizer is purchased from retail shops in Chiang Mai, or from village dealers who extend credit to the end of the garlic harvest for an interest charge of 20-40 percent. The fertilizers used are manufactured by American or Japanese firms.

Preparation of Seed

Garlic produced and saved the previous year is broken by hand into individual cloves. When planted, each clove is capable of producing an entire head. Large heads are set aside for use as seed the following year. Small children often participate in the work of *taek haum* (breaking up the heads of garlic). Neighbor children sometimes are paid to do the work at a standard rate of 0.25 *baht* for each two liters of garlic cloves produced.

Plowing

The field is plowed once. If buffalo and plow are used, the soil is first irrigated.

Preparation of Beds and Ditches

After plowing, the soil is allowed to dry. Then, with a heavy hoe, the field is formed into alternate planting beds and irrigation ditches. A light hoe is then used to break the clods on the beds.

Planting and Preparation of Straw

Planting begins when the soil is thoroughly dried. Individual garlic cloves are planted at 15 cm. intervals. The root side is placed down, and planting is very shallow, so that the top of the clove extends above the surface.

At the same time, bundles of straw are carried to the field. Many farmers acquire straw from their own plantings of early or late rice. Landlords often receive straw as part of a rental arrangement. There is also a cash trade in the commodity. In late fall, 1969, 100 bundles of rice straw (sufficient mulch to cover one *rai* of garlic) brought 30 *baht*. The previous year much more garlic had been planted, and the established price was 50 *baht* per 100 bundles. Late rice straw is considered superior because it is relatively long, but early rice straw is also used.

Application of Fertilizer and Straw

After the garlic cloves are in place, fertilizer is spread by hand. For this first application *puj daeng* (N-P-K) is most often used. After fertilizing, three to five centimeters of straw are applied to the beds.

Irrigation

A day after the garlic is planted, it is irrigated and watered for the first time. A bund is broken temporarily so that the small ditches can be filled. Water is then transferred from ditch to straw-covered bed with a woven bamboo scoop. Watering is carried out five or six additional times, at fifteen-day intervals, before harvest.

Weeding

After about one month, grasses which sprout in the bed are removed by hand. If weed growth is heavy, it is sometimes necessary to repeat this task after another month. After the first weeding, additional fertilizer is sometimes applied. Usually the fertilizer used is *puj khaaw* (urea) or a mixture of *puj khaaw* and *puj daeng.*

Harvest, Carrying from Field, Preparation for Sale, and Selling.

After about three months, the garlic is ready to harvest. This is done by pulling the dried plants. Tasks following harvest depend, to some extent, on the sales method chosen. If the garlic is to be sold fresh, the tops of the plant are cut in the field. Garlic stems are cut about four inches above the bulb. The garlic is then carried to the farmer's home, where the roots are also removed, the heads are sorted into large and small sizes, and the garlic is carefully washed.

Fresh garlic is sold to dealers who come to the village at this season (late February and early March) for this specific purpose. They come each afternoon for two or three weeks, with their trucks and platform scales, and set up business in front of the village temple. The dealers are middlemen, representing factories in Bangkok and other cities in central Thailand, which produce large quantities of pickled garlic. The price fluctuates from day to day, and, of course, from year to year. Price is to some extent negotiable, although the dealers offer a set price each day. The following case provides some insight:

> Two local girls came to sell their garlic to the dealer from Rim Ping (a village to the south of Sansai). They did not bother to inquire about the price being offered that day by the other dealer (from Lamphum). (In later discussion it came to light that they previously had done business with the Rim Ping middleman). The girls were offered 31 *baht* for their garlic. They pointed out that in the past they would have expected at least 50 *baht* for what they brought. They asked 40 *baht*. The woman responsible for doing the weighing said that they would have to ask the man in charge. 'The girls could offer their garlic to the other dealer present, but he did not give a fair price, etc.' Eventually he offered 35 *baht* and the girls accepted.

Drying

In deciding whether to sell fresh or dried garlic, the farmer must anticipate the price to be received for dried garlic several weeks or months in the future. He must take into account the probability of substantial annual price fluctuation, and the fact that his garlic will lose half its weight in the drying process. The following data on price fluctuation were gathered:

1967 - Fresh garlic - Large head = 4 *baht*/kg.
Small head = 3 *baht*/kg.

Dried garlic - Large head = 12 *baht*/kg.
Small head = 6-7 *baht*/kg.

1968 - Fresh garlic - Large head = 1.20 *baht*/kg.
Small head = 0.50 *baht*/kg.

Dried garlic - Large head = 2 *baht*/kg.
Small head = 1 *baht*/kg.

1969 - Fresh garlic - Large head = 2.50 *baht*/kg.
Small head = 1-1.50 *baht*/kg.

Dried garlic - Large head = 4-5 *baht*/kg.
Small head = 2-3 *baht*/kg.

1970 - Fresh garlic - Large head = 1.20-1.50 *baht*/kg.
Small head = 0.50 *baht*/kg.

It is clear that price fluctuations are substantial. Selling dried garlic is the best strategy in some, but not all, years. Dried garlic can be sold to a middleman at any time of the year. Well-to-do farmers are able to wait for many months, if necessary, until the price rises.

Garlic which is to be dried, or used for seed, is left in the ground for 10 or 15 days longer than that which is to be sold fresh. After the garlic is pulled, it is carried to the farmer's house. It is spread on mats for several days of drying. It is turned or stirred at regular intervals to insure even drying. The garlic is not exposed to direct sunlight, but dried under a house or a rice storage bin.

Bundling

After drying, and before storage, garlic is tied in bundles. The dried stems are tied together to form bundles, the characteristic unit in which garlic is wholesaled and retailed in the area.

Storage

Bamboo racks are used to suspend dried garlic bundles from the rafters or floor beams of the house or rice storage bin.

Decision Rules for the Garlic Crop

Garlic is an *expensive* crop which requires substantial inputs of fertilizer and labor. It is also a *gambler's* crop, subject to large, irregular, and annual price fluctuations. Through quantitative analysis, use of field/crop maps, and discussions with Sansai farmers, the following four decision "rules" have been explicated for the garlic crop at Sansai.

First, on the average, production, processing, and sale of the garlic crop requires 60-65 man-days of labor per *rai*. (See Table 24.) This in-

Table 24 *Labor Expenditure for the Garlic Crop, 1969-70*

	Operation	Mean Expended (Man-days/*rai*)	(SD)
1.	Buy fertilizer	0.41	(1.05)
2.	Preparing garlic	8.68	(7.27)
3.	Plowing	0.91	(2.28)
4.	Dig ditches and prepare beds	6.99	(6.07)
5.	Plant and prepare straw	15.37	(7.30)
6.	Laying fertilizer and straw	1.15	(1.11)
7.	Irrigation	5.75	(5.18)
8.	Weeding	3.42	(5.15)
9.	Harvest, cutting, carrying, selling	16.62	(11.74)
10.	Drying	0.25	(0.97)
11.	Bundling	2.78	(4.82)
12.	Storing	0.20	(0.37)
	TOTAL (N = 79)	62.53 Man-days	

cludes a total of 12 possible tasks, some of which are deleted if the garlic is sold fresh, others if the garlic is sold after drying. While the mean labor input is 62.5 man-days per *rai*, inputs as low as 53 man-days and as high as 80 man-days are well within the ordinary range. (See Table 25.)

Second, suitable garlic beds can be made in virtually all of Sansai's fields. (See Maps 14, 15, and 16.) The small ditches between beds can be varied in depth to assure an adequate supply of gravity-delivered irrigation water. However, garlic cannot be raised on fields which are separated from irrigation sources by watermelon fields. The latter must remain relatively dry; no irrigation water can enter them.

Third, for the garlic crop in Sansai, 1969/70 mean yields were significantly related to: (a) landholding status of the farming household; (b) amount of fertilizer applied to the land; and (c) total labor inputs. (See Tables 25, 26, and 27.)

In 1969/70, owning households achieved significantly higher yields than renting households. This probably reflects the opportunity that owners

Table 25 *Relationship Between Yield and Labor Expended on Various Tasks for the Garlic Crop, 1969-70*

		Low Input (N = 31)	(SD)	Medium Input (N = 36)	(SD)	High Input (N = 15)	(SD)
1.	Buy fertilizer	0.20	(0.18)	0.50	(1.32)	0.72	(1.25)
2.	Preparing garlic	8.37	(5.63)	7.10	(4.65)	12.01	(12.23)
3.	Plowing	0.54	(0.76)	0.84	(1.50)	1.79	(4.45)
4.	Dig ditches and prepare beds	6.66	(6.20)	6.49	(5.23)	8.22	(7.04)
5.	Plant and prepare straw	13.34	(7.01)	16.63	(7.54)	15.47	(6.54)
6.	Laying fertilizer and straw	0.87	(0.54)	1.39	(1.47)	1.09	(0.67)
7.	Irrigation	4.40	(4.50)	5.91	(5.03)	7.75	(5.68)
8.	Weeding	2.34	(3.50)	3.31	(5.92)	5.90	(5.10)
9.	Harvest, cutting, carrying, selling	13.84	(10.59)	15.34	(9.95)	24.11	(13.92)
10.	Drying	0.09	(0.22)	0.39	(1.37)	0.19	(0.50)
11.	Bundling	2.41	(5.71)	2.84	(4.16)	2.83	(3.81)
12.	Storing	0.10	(0.21)	0.32	(0.47)	0.06	(0.14)
	TOTAL	53.16		61.06		80.14	
	YIELD	418 kg/*rai*		759 kg/*rai*		1440 kg/*rai*	

Map 14 *Garden Crop Choice at Sansai, 1967-68*

Map 15 *Garden Crop Choice at Sansai, 1968-69*

Map 16 *Garden Crop Choice at Sansai, 1969-70*

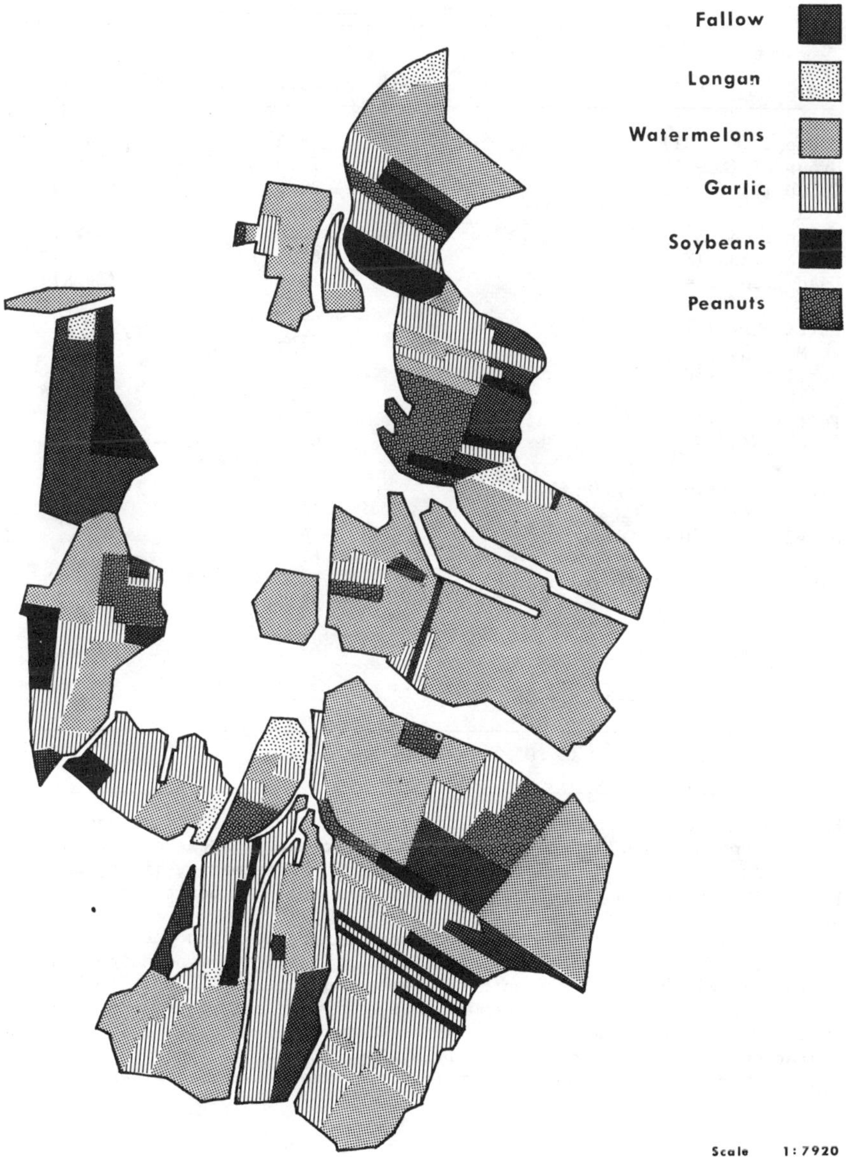

Table 26 *Yield Figures Under Various Conditions For the Garlic Crop, 1969-70*

	Mean Yield (10 kg. Units/*Rai*)	(SD)
Group I (N = 31)	77.2	(38.4)
Group II (N = 28)	73.0	(36.9)
Group III (N = 20)	79.0	(67.9)
Owned land (N = 35)	84.6	(58.3)
Rented land (N = 40)	65.1	(31.2)*
Used free (N = 4)	119.8	(24.9)
Fertilizer (8-79 *bhat*/*rai*) (N = 33)	66.7	(58.7)
Fertilizer (80-160 *baht*/*rai*) (N = 35)	69.2	(27.7)*
Fertilizer (161-350 *baht*/*rai*) (N = 34)	102.5	(38.4)
Hired labor (0-9 days/*rai*) (N = 34)	72.1	(45.1)
Hired labor (10-19 days/*rai*) (N = 36)	67.9	(30.4)
Hired labor (20-68 days/*rai*) (N = 17)	76.1	(77.5)
Sold in field (N = 13)	69.2	(24.5)
Harvested, sold fresh (N = 65)	77.1	(50.9)

* Significant at .05 level.

have to choose land which is especially well suited to garlic production. Plots planted in garlic are seldom more than two *rai*. Owners of rather small plots are able to rent out a *rai* or two. Presumably, it is land of poorer quality which is rented out, i.e. land which is not easily irrigated.

The difference in yield between owned and rented land does not reflect significant variation in fertilizer application. Owners spent an average of 117 *baht* per *rai* on fertilizer, while renters spent 110 *baht*.

Yield figures (1969/70) demonstrate a significant relationship to amount of fertilizer applied. In particular, *very heavy* fertilizer application results in correspondingly high yield. The following relationship between fertilizer cost and mean yield exists (see below Table 27).

Table 27 *Sales Figures Under Various Conditions For the Garlic Crop, 1969-70*

	Mean Sale Price (*Baht/Rai*)	(SD)
Group I (N = 31)	891	(441)
Group II (N = 28)	751	(407)
Group III (N = 20)	773	(427)
Owned land (N = 35)	826	(419)
Rented land (N = 40)	766	(388)
Used free (N = 4)	1258	(530)
Fertilizer (8-79 *baht/rai*) (N = 33)	584	(338)
Fertilizer (80-160 *baht/rai*) (N = 35)	819	(350)*
Fertilizer (161-350 *baht/rai*) (N = 19)	1155	(444)
Hired labor (0-9 days/*rai*) (N = 34)	821	(470)
Hired labor (10-19 days/*rai*) (N = 36)	769	(380)
Hired labor (20-68 days/*rai*) (N = 17)	698	(462)
Sold in field (N = 13)	954	(360)
Harvested, sold fresh (n = 65)	778	(431)
Low Labor Input (N = 31)	545	(238)
Medium Labor Input (N = 36)	822	(359)*
High Labor Input (N = 15)	1281	(447)

* Significant at .01 level

Fertilizer Cost (*baht/rai*)	Increment[1] (kg./*baht* of fertilizer)
8-79 (N = 33)	4.8
80-160 (N = 35)	3.5
161-350 (N = 19)	3.7

The cost effectiveness of fertilizer declines with heavier application. However, heavy fertilizer application is easily justified in terms of expected profit:

Fertilizer Cost *(baht/rai)*	Sale Price *(baht/rai)*	Total Expenditure[2] *(baht/rai)*	Expected Profit *(baht/rai)*
8-79 (N = 33)	584	107	477
80-160 (N = 35)	819	220	599
161-350 (N = 19)	1155	316	839

Households in 1969/70 receiving low yields devoted smaller amounts of labor to garlic production than households receiving higher yields. In part, variation in labor inputs is a *result* of differential yield. For example, tasks associated with harvest (task #9) were more time-consuming for households with high yields. It is reasonable to assume that it takes longer to harvest more produce. However, labor inputs into certain tasks can best be considered to *cause* variation in yield. Labor inputs into tasks (1) to (8), all carried out before harvest, are as follows:

Yield (kg./*rai*)	Labor Expenditure (man-days /*rai*)
Households receiving yield of 150-590 kg/*rai*	36.7
Households receiving yield of 600-1000 kg/*rai*	42.2
Households receiving yield of 1010-2400 kg/*rai*	53.0

These figures indicate a probable causal relationship between labor inputs before harvest and yield.

Fourth, each year Sansai garlic farmers must store some of their produce to provide seed for the coming year. The ratio of seed to harvested garlic is on the order of 1:5 or 1:10, so that the amount stored is necessarily large. Sales figures do not reflect yield, but rather the size of the farmer's surplus beyond his projected seed requirements for the coming year. Nevertheless, sales figures are significantly related to fertilizer use and labor inputs.

Before choosing to raise garlic during the garden crop season at Sansai, the farmer is likely to consider the following four factors. The first is *seed supply*. If he did not raise garlic the previous year, he will not have any seed on hand. It will be necessary to buy some at the prevailing price. The second is *labor inputs*. Garlic cultivation requires heavy labor expenditure. Households with large labor forces or ample cash reserves are in a strong position to raise this crop. Others are not. The third factor to be considered is *fertilizer cost*. Successful cultivation of garlic requires application of large amounts of chemical fertilizer. Well-to-do farmers can expect higher profits, because they can afford to buy more fertilizer. Poorer farmers, forced to buy on credit, will receive less fertilizer for their money. The fourth is *expected price*. Sansai farmers are aware that garlic prices have been subject to drastic fluctuations in recent years. For example, a crop that yielded

2800 *baht* in 1967 would have brought only 680 *baht* in 1968. Even at very low prices (1968 or 1970), farmers achieving high yields (more than 1200 kg. per *rai*) can expect a profit of several hundred *baht* per *rai*. However, farmers achieving low yields can easily lose money in a year when prices are low.

Watermelons

Watermelons are a new crop at Sansai. Their cultivation is technically complex and requires heavy inputs of fertilizer and labor. Melons are the only Sansai crop for which no permanently established middleman system has been established. Farmers who raise the crop must accept the difficult, time-consuming, and expensive task of transporting their produce to a nearby or distant city for retail sale. And yet, it is well known locally that in recent years some Sansai farmers have received high yields and profits from watermelons. A year of great expansion of watermelon cultivation in the Chiang Mai Plain occurred in 1970. The market was glutted, and farmers retailing their melons in Chiang Mai or Lamphun, or as far away as Lampang, received very low prices from consumers. Better prices were received by farmers able to ship their fruit as far as Nan, Mae Saaj (on the Burmese border), or Maehongson.

The watermelons raised at Sansai are of the small, "icebox" type. Ordinarily, commercial seeds, processed and canned in the United States, are used. A seed variety called *bang berd* (developed at Prince Sithiporn's farm in southern Thailand) is also planted.

Cultivation of the watermelon crop can be divided into fourteen major tasks: purchase of seed and fertilizer; plowing; digging wells; making seed cups; planting seeds; digging small holes; watering seedlings; planting seedlings and fertilizing; adding additional fertilizer; watering plants; weeding; harvest and grading; locating a truck; and transport and sale. Each task will be considered in turn.

Purchase of Seed and Fertilizer

Watermelon seed and commercial fertilizer are purchased in the city. Seed is ordinarily purchased in one pound cans which cost 60-100 *baht* each, depending upon brand and variety. One can provides sufficient seed for six or eight *rai*. Often, two or more farmers will together buy a single can. One of the most popular varieties at Sansai is "Sugar Baby." Fertilizers used include *puj daeng* (N-P-K), *puj khaaw* (urea), and chicken manure. All three often are used in combination. Chicken manure is purchased from a chicken farm near Chiang Mai at a price of two *baht* per five gallon can.

Plowing

Either draft animals or tractors are used in plowing. Standard price for the tractor at this season is 80 *baht* per *rai*. The tractor, of course, operates best on dry soil.

Digging Wells

At Sansai watermelons are watered by hand from shallow wells. One well, perhaps 2.0 meters long, 1.5 meters wide, and 1.5 to 2.0 meters deep, is sufficient to water two or three *rai* of melons. Digging is done with a heavy hoe.

Making Seed Cups

Watermelon seeds are not planted directly into the soil. Instead, seeds are planted with a small amount of soil in a "cup" or "bowl" formed out of banana leaves, or sometimes, in a small plastic bag. Each *rai* of melons requires about 1000 banana leaf cups.

Planting Seeds

When all the containers have been assembled, they are carried to the field and filled with soil. Two or three seeds are then placed in each container.

Digging Small Holes

Seedlings eventually are placed in small holes dug with a light hoe. The approximate dimensions of each hole are those of a shallow, upside-down cone approximately 45 cm. in diameter at the top, and 40 cm. deep. Laborers hired to do this task sometimes are paid on a piece-work basis (0.14-0.15 *baht* per hole).

Watering Seedlings

After seeds are placed in the banana leaf containers, the containers are placed compactly, in long rows, on the ground. They are watered each day until they are transplanted, after about 15 days. Water is drawn from the shallow wells with a large watering can. Several thousand seedlings can be watered in a matter of minutes.

Planting Seedlings and Adding Fertilizer

The next step is to add a few handsful of dirt and chicken manure or chemical fertilizer to each small hole and then stir. The hole is filled to within 10 cm. of the top. At the time of transplanting, watermelon seedlings are about 15 cm. tall. The seedlings are removed from their containers, placed in the center of the hole (on top of the soil/fertilizer mixture), and the hole is filled with soil. A handful of chemical fertilizer is applied around each plant seven days later. At this point, *puj daeng* fertilizer often is used.

About fourteen days after planting, additional fertilizer is applied. At this point, urea, mixed with water, often is the choice.

Watering Plants and Weeding

The most time-consuming task in watermelon cultivation is hand watering of the plants. The worker wears a yoke across his shoulders from which are suspended two large watering cans. Each plant is watered individually with water drawn from the shallow wells. Watering is done every second or third day, every second day if the soil is sandy. Although some hand weeding is done, relatively little time is devoted to this task.

Harvest and Grading

About seventy-five days after transplanting, the watermelons are ripe. Farmers judge ripeness by tapping the fruit with a knuckle of the index finger. A small knife is used to sever each fruit from the vine. Usually the fruits are sorted into "small," "medium," and "large" sizes in the field.

During the few weeks before harvest there is danger that nearly ripe watermelons will be stolen. In consequence, one member of the cutivating household ordinarily sleeps in the field in a small grass hut. This is not considered as a "task" of watermelon cultivation, but it is a considerable inconvenience to the household.

Finding a Truck

During March there are hundreds of watermelon farmers located within a few kilometers of Sansai who require transport for their crop to the market. It is not surprising that truck rental is expensive at this time. Some farmers require two or three days to arrange transport. Trucks from the area are in almost continuous operation. Truck drivers, spending as much as 12-16 hours each day behind the wheel, frequently are involved in accidents.

Transport and Sale

Sansai farmers attempt to retail their watermelons as near to the village as Saraphi and as far away as Nan, Mae Saaj, and Maehongson. The most common choice is Lampang, a three hour truck ride from the village. At Lampang, Sansai farmers set up a series of bamboo and canvas stands side-by-side along a major road entering the city. Well-to-do farmers sometimes hire village women to sell their watermelons.

Decision Rules for the Watermelon Crop

Watermelons are an expensive crop that requires heavy inputs of fertilizer and labor, and also involve purchased seeds, and sometimes insecticide. The sale of watermelons is a particularly onerous task. It is the only crop raised at Sansai which cannot be easily sold to a middleman. In 1970, in response to a market glut, Sansai watermelon farmers received very low prices.

Through use of quantitative analysis, reference to field/crop maps, and discussions with Sansai farmers, the following four decision "rules" have been explicated for the watermelon crop at Sansai.

First, on the average, production and sale of the Sansai watermelon crop requires 55-65 man-days of labor for each *rai* planted. (See Tables 28 and 29.) This includes a total of fourteen major tasks.

Second, watermelons can be raised in virtually all Sansai fields during the garden crop season. (See Charts 14, 15, and 16.) The village's lowest plots must be avoided only if adjoining plots are to be planted in rice before the watermelons are harvested. In this case, there is danger that maturing melons will rot due to excessive moisture resulting from seepage. Under these conditions, Sansai's farmers are influenced by decisions made for neighboring plots.

Table 28 *Labor Expenditure on the Watermelon Crop, 1969-70*

	Mean Expended (Man-days/*rai*)	(SD)
1. Buying seed, fertilizer	0.31	(0.42)
2. Plowing	0.73	(1.47)
3. Digging wells	1.11	(1.16)
4. Making seed cups	3.51	(2.96)
5. Planting seeds	2.17	(2.05)
6. Digging small holes	7.32	(6.47)
7. Watering seedlings	1.66	(2.34)
8. Placing fertilizer and seedlings	7.55	(4.55)
9. Add additional fertilizer	6.94	(5.65)
10. Water plants	17.71	(6.43)
11. Weeding	0.91	(2.24)
12. Harvest/grading	5.91	(6.30)
13. Locating truck	0.43	(0.78)
14. Transporting/selling	3.14	(2.71)
TOTAL (N = 54)	59.40 Man-days	

Third, for the watermelon crop at Sansai, 1969/70 mean yields were significantly related to quantity of fertilizer used[3] and use or non-use of a tractor. (See Table 30.) In particular, very heavy fertilizer application results in correspondingly high yield. In 1969/70, the following relationship existed between fertilizer cost and yield:

Fertilizer Cost (*baht/rai*)	Increment (Watermelons/*baht* of Fertilizer)
50-149 (N = 16)	1.3
150-250 (N = 20)	2.0
251-500 (N = 16)	2.8

Table 29 *Relationship Between Yield and Labor Expended on Various Tasks for the Watermelon Crop, 1969-70*

	Low Input (N = 20)	(SD)	Medium Input (N = 15)	(SD)	High Input (N = 6)	(SD)
1. Buying seed, fertilizer	0.16	(0.19)	0.43	(0.50)	0.48	(0.69)
2. Plowing	0.68	(1.73)	0.81	(1.21)	1.33	(1.70)
3. Digging wells	1.09	(1.62)	1.19	(0.80)	0.82	(0.61)
4. Making seed cups	3.46	(2.43)	3.86	(4.55)	2.90	(1.01)
5. Planting seeds	2.17	(1.44)	2.74	(3.03)	1.93	(1.16)
6. Digging small holes	6.03	(3.71)	8.25	(8.95)	9.50	(6.69)
7. Watering seedlings	1.88	(3.26)	2.16	(2.03)	1.35	(0.78)
8. Placing fertilizer and seedlings	8.00	(4.18)	5.81	(3.32)	6.83	(5.01)
9. Add additional fertilizer	6.77	(4.04)	5.73	(3.19)	6.87	(3.00)
10. Water plants	16.49	(6.47)	19.73	(5.13)	18.27	(6.34)
11. Weeding	1.13	(2.71)	1.09	(2.50)	0.67	(1.49)
12. Harvest/grading	4.57	(4.64)	5.03	(2.31)	10.03	(12.07)
13. Locating truck	0.27	(0.40)	0.79	(1.27)	0.33	(0.34)
14. Transporting/selling	2.62	(2.28)	3.43	(2.98)	4.20	(2.14)
Total Labor Expended	55.32		61.05		65.51	
	631 melons/ *rai*		1190 melons/ *rai*		2110 melons/ *rai*	

Table 30 *Yield Figures Under Various Conditions for the Watermelon Crop, 1969-70*

	Mean Yield (Fruits/*Rai*)	(SD)
Group I (N = 15)	1164	(724)
Group II (N = 19)	970	(473)
Group III (N = 5)	864	(299)
Owned land (N = 15)	819	(374)
Rented land (N = 12)	1051	(557)
Use free (N = 6)	1005	(515)
Tractor used (N = 16)	1143	(618)
Tractor not used (N = 14)	874	(417)
Fertilizer 50-149 *baht/rai* (N = 16)	814	(332)
Fertilizer 150- 250 *baht/rai* (N = 13)	1008	(660)*
Fertilizer 251-500 *baht/rai* (N = 10)	1408	(586)
Insecticide 0-59 *baht/rai* (N = 10)	1092	(540)
Insecticide 60-100 *baht/rai* (N = 22)	1062	(656)
Insecticide 101-200 *baht/rai* (N = 9)	978	(328)
Sold Retail (N = 33)	1103	(618)
Sold Wholesale (N = 8)	835	(207)

* Significant at .05 level

The cost effectiveness of fertilizer declines rapidly with heavy application. In the case of the watermelon crop (unlike that of garlic) very heavy fertilizer application cannot be justified in terms of expected profit.

Fertilizer Cost (*baht/rai*)	Sale Price (*baht/rai*)	Total Expenditure[4] (*baht/rai*)	Expected Profit (*baht/rai*)
50-149 (N = 16)	493	307	186
150-250 (N = 20)	787	367	420
251-500 (N = 8)	909	692	217

Higher yield is associated not only with higher fertilizer expenditures, but also with much higher transport costs.

Tractor plowed fields are, for some undetermined reason, agronomically superior to animal plowed fields. Perhaps the tractor, plowing deeper on a field leaves the soil drier and in a condition more appropriate for melon growth.

Fourth, because yield values expressed in number of watermelons are not especially accurate, it is worthwhile also to give consideration to the "price received." (See Table 31.) Sales figures for 1969/70 are significantly related to the use or non-use of tractor, method of sale, and amount of fertilizer used.

Table 31 *Sales Figures Under Various Conditions for the Watermelon Crop, 1969-70*

	Mean Sales Value (10 *baht* Units/*Rai*)	(SD)
Group I (N = 19)	73.6	(30.4)
Group II (N = 20)	77.0	(44.2)
Group III (N = 13)	67.5	(37.9)
Owned land (N = 19)	52.1	(28.1)
Rented land (N = 16)	79.9	(41.2)
Used free (N = 9)	88.9	(33.2)
Tractor used (N = 39)	81.1	(38.8)
Tractor not used (N = 15)	62.1	(38.7)**
Fertilizer 50-149 *baht/rai* (N = 16)	49.3	(26.8)
Fertilizer 150-250 *baht/rai* (N = 20)	78.7	(34.5)*
Fertilizer 251-500 *baht/rai* (N = 8)	90.9	(40.3)
Insecticide 0-59 *baht/rai* (N = 14)	88.4	(40.5)
Insecticide 60-100 *baht/rai* (N = 28)	70.3	(36.1)
Insecticide 101-200 *baht/rai* (N = 12)	74.2	(43.3)
Sold Retail (N = 43)	80.7	(40.6)
Sold Wholesale (N = 11)	56.6	(28.9)*

* Significant at .05 level

* Significant at .10 level

The few farmers who choose to sell their watermelons *in* the village (to other farmers also engaged in selling their own melons) probably do so for two reasons. First, their plots are, on the average, very small--only 0.54 *rai*, as against 1.91 *rai* for those selling *outside* the village. For such small plots potential profits are very small in proportion to the extra expense and labor involved in retailing the melons. Second, those households selling within the village achieved rather low yields. Again, potential profits must seem low in proportion to the expense and labor involved in retailing the watermelons.

It is clear that households applying greater amounts of fertilizer achieved both higher yield and sales. However, Table 31 does demonstrate that increased expenditures on fertilizer are associated with increased expenditures in other areas as well, e.g. labor and transport. If we consider expenditures on fertilizer, labor, and transport, some high-yielding operations appear to be "over-financed" or inefficient. Yield and sale price continue to increase with increasing cash expenditures, but at some point expenditures begin to increase faster than the yield.

Peanuts

Peanuts can be raised on any Sansai plot. They are raised on narrow, heaped-up strips of soil, with small irrigation ditches between. Peanuts require nearly as much labor as watermelons or garlic, and moderate cash expenditures--for seed, and in some cases for tractor rental and hired labor. Cultivation of the peanut crop can be divided into ten major tasks, as follows: purchase and preparation of seed; plowing; construction of earth strips and ditches; planting; irrigation; weeding; harvest; carrying peanuts from field; drying and preparation for sale; and sale. Each of these tasks is discussed in turn.

Purchase and Preparation of Seed

Peanut seed is purchased from a dealer in Sansai or a nearby village. A bucket of unshelled peanuts for planting costs 16-18 *baht*. Two buckets are sufficient to plant one *rai*. Several days before planting the peanuts are shelled by hand. Three nights before planting the seeds are soaked overnight. The next day they are placed in a basket lined with papaya leaves and covered with leaves. After two nights the sprouting peanuts are ready for planting.

Plowing and Construction of Earth Strips and Ditches

The field is plowed once, either by tractor or draft animal. A heavy hoe is used to form the soil into a series of ridges and ditches. Each ridge is 40-50 cm. wide, and level on top. This task requires four or five man-days of labor per *rai*.

Planting

Before planting the peanuts, the small ditches between ridges are filled nearly to the top with irrigation water. After planting, the water level is lowered so that the ditches are half full. After initial flooding, a series of small holes are made along each ridge. One stroke of the hoe is sufficient to form the holes that are spaced at 25 cm. intervals. Planting consists of placing four seeds (shelled peanuts) in each hole and covering them with dirt. The entire task requires about eight man-days of labor per *rai*.

Irrigation

The water left standing in the small ditches is eventually lost to evaporation and by absorption into the adjoining ridges. After one month, the water is raised again until the ditches are half full. If the water supply is sufficient, this process is repeated after two and three months. There is seldom enough water for the third month.

Weeding, Harvest, and Carrying Peanuts from the Field

Before first irrigation, grass and weeds are pulled by hand. This task requires about three man-days of labor per *rai*. About four months after planting, when the visible leaves are dry and brown, the peanuts are ready for harvest. They are pulled by hand and adhering soil is knocked off against the foot or leg. Harvest requires 10-15 man-days per *rai*. The peanuts are then carried from the field to the house in carrying baskets.

Drying and Preparation for Sale

The peanuts are then prepared for sale. They may be sold either as *thua din haeng* (dried peanuts) or *thua din dip* (fresh, boiled peanuts). Dried peanuts are prepared by carefully separating the peanuts (in the shell) from all foliage, and then drying them on mats in the sun. Total labor required is on the order of 20-30 man-days per *rai*. After cleaning, peanuts to be sold fresh are boiled, and sold immediately. Dried peanuts may be held as long as a year before sale.

Sale

Whether peanuts are sold fresh or dried, they can be sold to middlemen. The middleman often seeks out the farmer and offers a price.

Decision-making for the Peanut Crop

In 1969/70 few Sansai farmers raised peanuts. In consequence, there are too few data available to construct decision "rules" which attempt to relate yield to other variables. However, mean values for the following items have been calculated: labor inputs; cash expenditure; and expected sale value.

Mean Labor Input = 55.0 man-days per *rai*. (N = 8)

The peanut crop requires labor inputs nearly as heavy as those required by garlic and watermelons.

Mean Expenditure = 136 *baht* per *rai*. (N = 8)

Production costs for the peanut crop include seed, tractor rental, and hired labor. Becuase no fertilizer is used, production costs are much lower than for garlic or watermelons.

Mean Expected Sale Value = 326 *baht* per *rai*. (N = 8)

Expected profits are on the order of 190 *baht* per *rai*, or less than four *baht* for each man-day of labor.

Soybeans

The cost of raising soybeans at Sansai is low, both in terms of labor inputs and production expenditure. The crop can be raised on any Sansai plot, and does quite well on low, damp soils. The farmer often is able to omit any plowing. Many farmers sell the entire crop before harvest (*khaaj maaw*) to temporary middleman/harvest groups which are formed locally. By doing so, they roughly cut in half their labor inputs. Soybean production can be divided into nine major tasks; purchase of seed; plowing; planting; irrigation; weeding; harvest; carrying soybeans from field; preparation of soybeans for sale; and sale. Each of these tasks will be discussed in turn.

Purchase of Seed

Soybean seed is purchased in Chiang Mai and various small towns. A bucket of seed, costing 40-45 *baht*, is sufficient to plant one *rai*. The seed is not processed in any way before planting.

Plowing and Planting

The plot may be plowed either by tractor or draft animal, but this operation is often unnecessary. Instead, rice stalks are cut near ground level with a knife, and the plot is ready for planting. If the soil of the plot is dry, it is dampened by throwing canal water on it with a *cun* (woven water scoop, also used for watering garlic). Small holes about 2.5 cm. deep and 40-45 cm. apart are made with the point of a stick. Five soybeans are placed in each hole. No additional labor is required until the plants are ready for harvest. A few farmers occasionally water and weed their soybeans.

Harvest

Many, perhaps most, Sansai soybean farmers sell their crops before harvest to cooperative harvest groups. These groups are made up of five or six individuals, usually middle-aged women, who cooperate in buying and harvesting an entire plot. The beans are pulled off the plants, and made into small bundles in the field. The bundles are divided evenly, and the women carry them to their homes. Each woman boils her beans and sells them in a retail market. At this point, members of the cooperative group are widely dispersed. Sansai women sell boiled soybeans at several markets in Chiang Mai, at San Pratong, and at several closer markets.

Decision-Making for the Soybean Crop

In 1969/70, few Sansai farmers raised soybeans. As a result, insufficient data are available to construct decision "rules" which relate yield to other variables. However, mean values for the following items have been calculated: labor inputs; production cost; and expected sale value.

Mean Labor Inputs = 16.8 man-days per *rai*. (Before harvest = 7.3; harvest and after = 9.5) (N = 11)

The labor required to produce a crop of soybeans is much lower than for peanuts, garlic, or watermelons. Farmers who sell their entire crop before harvest devote as little as six or eight days of labor for each *rai* cultivated.

Mean Production Cost = 60.0 *baht* per *rai*. (N = 11)

The major cash expense in soybean production is the price of seed. Some farmers also spend cash for tractor rental and hired labor.

Mean Expected Sale Value = 119 *baht* per *rai*. (N = 11)

On the average, the Sansai soybean farmer can expect to clear about 60 *baht* for each *rai* cultivated--i.e. 119 *baht* sales - 60 *baht* production cost. Labor inputs are slight, but expected profit is proportionally low.

Choosing a Garden Crop

Chart 17 presents the decision rules discussed earlier in this Chapter. The insights to be gained are similar to those provided for early and late rice in Chapter Five. The chart over-simplifies an essentially complex situation. Sansai residents choose among four different garden crops. However, previous decisions (e.g. to plant late rice after August floods) may effectively eliminate some choices. For example, garlic and watermelons cannot be effectively raised after late harvested late rice. Relative costs are also crucial; watermelons and garlic are relatively expensive crops, while peanuts and soybeans are relatively inexpensive.

Garden crops can easily be planted in combination, and in various amounts. Sansai households are really faced with many more choices than four. In theory, possible choices are infinite.

The household choosing a crop or combination either owns or rents the land used. This is not ordinarily a matter of "choice," but a matter of "fact," and is so indicated by a broken line in Chart 17. The nature of the rental arrangement varies so the renting household may exert some control. Some households pay cash rent (usually 100 *baht* per *rai*), while others work under a share-cropping (half-shares on expenses and production) arrangement. This is a matter for negotiation with the landlord.

Garden crops often are expensive, and cash losses are known. Some landlords are willing to enter into risky, but potentially profitable, share-cropping arrangements. Others prefer the security of a cash rent. Likewise, some renting households prefer the former and others the latter arrangement. Sharecropping will be attractive to poor households unable to raise sufficient cash for fertilizer, seed, and other expenses. Cash rental holds out the possibility of both higher profit and greater loss. The lines in Chart 17 representing "cash rental" and "sharecropping" are dashed to indicate that rental arrangement is to some extent negotiable.

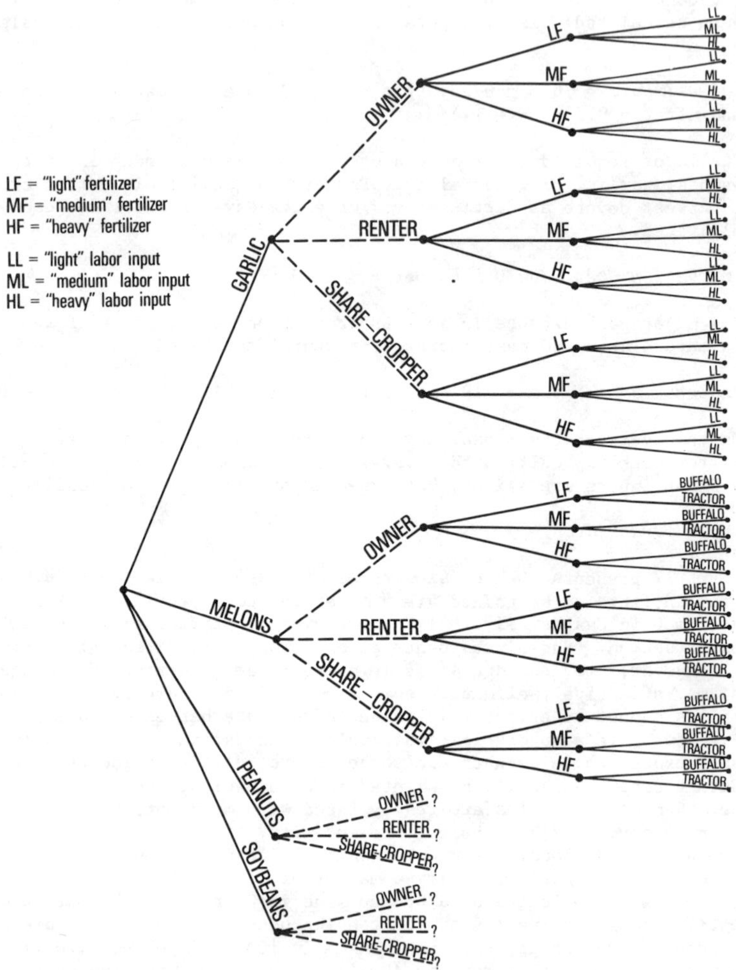

Chart 17 *Decision Tree for Garden Crops*

After choice of garlic, the farming household is faced with two other significant choices: amount of fertilizer and amount of labor. Both choices are subject to gradation, so that "yield" can best be expressed as a linear function of the two. Choice of watermelons is also followed by two significant choices: amount of fertilizer and method of plowing.

Since the former variable is subject to infinite gradation, the expected yield can best be expressed as a linear function of fertilizer inputs. The latter variable involves a discrete choice between tractor and buffalo; the expected yield is expressed as a discontinuous function of plowing method.

Households raising soybeans or peanuts probably are faced with choices which have significant effect on yield and profit. However, too few farmers raised these crops in 1969/70 to provide an adequate test.

Strategies in raising garden crops will not receive further discussion in this monograph. However, it is assumed that a slightly more complex version of the model developed in Chapter Seven would be appropriate to description of choosing a garden crop at Sansai.

Chapter Seven

A TENTATIVE MODEL OF CROP CHOICE AT SANSAI

In economic anthropology, agricultural economics, and related fields, the specific nature of "peasant," "primitive," or "subsistence" decision-makers has been at issue for several decades. Some scholars argue that technologically simple farmers are unable or unwilling to respond to material stimuli. They claim that these individuals can be expected in many situations to maximize non-material goals. Others argue that these farmers are no less responsive to material stimuli than Western farmers. They differ from the latter only to the extent that they are poorer, cannot afford elaborate capital inputs, and operate in systems where production factors have very different relative prices. They claim that these farmers can be expected to maximize material goals in most situations. In summary, the debate focuses on whether farmers who live in developing countries and produce crops with simple technology are essentially material or non-material maximizers.

This debate is of more than academic interest. If farmers are material maximizers, their behavior should be interpreted against background data on yields, production costs, and profits. If they are non-material maximizers, their behavior should be interpreted in relationship to needs for social solidarity, maintenance of value systems, and conformity to existing cognitive structures. It is reasonable to assume that farmers of each type should be approached differently in economic development programs. Material maximizers are primarily in need of research, information, and credit services. Non-material maximizers pose a far more complex problem.

Farmers Viewed as Non-Material Maximizers

Scholars who have emphasized the importance of non-material goals often are called "substantivists," apparently because they are concerned with studying the *substance* of non-Western economic systems, rather than imposing inappropriate interpretations derived from Western economic theory. Individuals who might usefully be assigned to this category include anthropologists (Foster, Nash, and Kingshill), economists (Dalton and Desai), and an economic historian (Polanyi).

Anthropologists often refer to Foster's notion of the "Image of Limited Good" as an example of the difficulties which development programs face (Foster 1965). This "Image" is associated with the notion that the world's supply of good things is finite. Therefore, a gain for any particular individual or group necessarily implies a loss for some other individual or group. Foster himself seems to waver on the precise implications of the "Image." On the one hand, he suggests that it represents a reasonable, self-interested reaction to traditional socioeconomic conditions. On the other hand, he implies that it constitutes a potential barrier to government development programs. For example, he assumes that the "Image" causes some farmers to reject programs that clearly promote their material self-interest.

Dalton states his position clearly:

> It is not so, as some literature of economic anthropology asserts, that the postulates of scarcity and economizing calculation are of universal relevance (1968:148).

He continues by arguing that

> Primitive economy is different from market industrialism not in degree but in kind. The absence of machine technology, pervasive market organization, and all-purpose money, plus the fact that economic transactions cannot be understood apart from social obligation, create, as it were, a non-Euclidean universe to which Western economic theory cannot be fruitfully applied (1968:164).

The intended inference is obvious: we are to believe that non-Western decision-makers are not motivated by material self-interest.

In his study of a village in Chiapas, Nash makes a subtly different point about Indian decision-makers:

> It is the contention of this paper that the rationale of economic choice in peasant society follows the same general rule of maximization as economic activity does anywhere, at any time. What is distinctive about peasant and primitive societies are not the habits of mind about advantage, nor an inability to calculate costs and benefits of a course of action, nor even an absence of a motive of gain; but rather the possession of a set of concrete social organizations which directly channel economic choice, on the one hand, and a set of sanctions which operate to keep economic deviants in physical as well as moral jeopardy on the other (1968:311).

That is, farmers using simple technology are motivated by material self-interest, but socioeconomic factors raise a significant barrier to the pursuit of these goals.

Several scholars have attributed a lack of material self-interest to Thai rice farmers. Kingshill completes a chapter on "The Village Economy" with the following paragraph:

> In general, it seems that the Ku Daeng villager can live a rather independent existence. He grows the rice he needs. He has his own vegetables in his garden. He raises his own meat supply and catches his own fish. Some of the things he needs are brought to him by merchants, some he buys in the market. His existence seems to be quite independent of the economy of

> the country as a whole. If the price of rice drops, he may not like it, and he may make a little less profit, but his basic means of existence will not have been changed. Not much will cause him to increase his efforts to earn a little more. If something strikes his fancy as worth having, and if it is only slightly beyond his means, he probably will increase his efforts. If it is, however, completely beyond his means, he will only dream about it, perhaps buy a lottery ticket, but never make any serious effort to increase his earning power (1965: 46).

Kingshill suggests that Ku Daeng farmers are not responsive to price shifts, and by implication, that they are motivated by non-material goals. A number of Thai scholars, according to Behrman, have attributed this quality to their rural countrymen (1968:7).

In most cases where scholars attribute non-material goals to technologically simple farmers, this condition is simply asserted, or based on anecdotal evidence. D. K. Desai's analysis of data from Maharashtra provides an interesting exception. His conclusions regarding the nature of decision-making by farmers are based on analysis of extensive quantitative data (1963).

In Desai's *Increasing Income and Production to Indian Farming* he applies linear programming techniques to evaluation of crop mixes on forty farms in two districts of Maharashtra State in India. Results obtained on each farm are judged against an "optimum" standard. The reader's first impression is that the nature of goals pursued by farmers in the studied group is not at issue. In general, explicit statements related to this matter are avoided. At a few points, however, positions are adopted which suggest belief that the farmers studied are significantly constrained in their pursuit of material goals. The following passage, stated under "Conclusions," reveals Desai's position:

> The results of the present investigation indicate a substantial potentiality of increasing farm income and production with the existing resource supplies and the present technical knowledge of the farmers. These results are based on the assumptions outlined in Chapter 1. Many people question the validity of the assumption of profit maximization motive of Indian farmers. It is argued that the social and cultural factors in rural India emphasize detachment and otherworldliness and the farmers are generally not interested in their material welfare because of the influence of these factors. There are many indications which can be put forth against this argument. Although there are indications of changes in social and cultural patterns of rural India, *there does not seem to be a conscious effort on the part of the farmers in general to bring about this change at a faster rate.* Unless the change in the motivation-pattern is brought at a faster rate, the rate of economic development is bound to be slow. Economic development involves a deep cultural and social change--a change in values, habits, knowledge, attitudes, ways of life, social ideals and aspirations.

> *This change cannot be expected to come about without special efforts of the enlightened section of the society.* As it should be, the Government of India is trying to help the farmers through the Five-Year Plans and Community Development Projects to bring about a change in their motivational pattern (1963:155, italics added).

Implications of the italicized portions of the preceding quotation seem apparent. The Indian farmers studied lack the psychological characteristics of materially self-interested decision-makers. They are not "enlightened" and are capable of reacting only slowly to advantageous change. It is in light of this conclusion that Desai's analytical techniques deserve to be carefully scrutinized.

Desai analyzes data collected in Maharashtra in 1956-57 as part of the Farm Management Investigation conducted by the Research Programmes Committee of the Planning Commission. The study

> examines the possibilities of increasing production and income on 40 individual farms, 20 each from the Ahmednagar and the Nasik district with the help of a linear-programming model. A comparison is made between the optimal and actual organizations and income levels of these farms (Desai, 1963:v).

Several of the procedures adopted by Desai deserve critical review. In the process, several of the intellectual difficulties involved in microanalysis of decision-making will be examined. Specific criticism can be leveled at Desai's use of: *ex post* price and yield values; district-wide mean price and yield values; and undue reliance on interviewer opinion.

In regard to the first criticism the following passage provides additional insight into analytical techniques adopted:

> In order to separate the problem of resource adjustment from its dynamic aspect with respect to time, we study it in the static framework, i.e., we assume the changes introduced and results obtained are instantaneous. This will enable us to use the convenient tools of economic theory suitable for this economic purpose (1963:3-4).

The major drawback to this approach is that it ignores one of the major features of agricultural decision-making in all parts of the world. Planting decisions are always made in the face of variable yields, and nearly always in the face of variable prices. The farmer must decide on the basis of *ex ante* judgments of these values. Desai ignores this aspect of decision-making by substituting *ex post* values. Adoption of this procedure ignores one of the most complex aspects of agricultural decision-making, and introduces a degree of vagueness and unreality. Given the procedures used, there is little apparent justification for labeling the choice behavior of the farmers studied as non-optimum. Because they are restricted to *ex ante* price and yield estimates, farmers are seldom able to choose ideal crop mixes. But Desai has offered no evidence that economists working under the same restrictions can do any better.

The second criticism of Desai's model is his assumption that each crop achieves uniform yields on all areas with similar soils, and that prices are uniform throughout each district. Once again, a major problem facing the rational decision-maker has been defined out of existence. Variability in yield and price must be dealt with. Desai ignores the possibility that crops of high mean yield and price may also be characterized by high variability, and thus by a high risk factor. Generally, he ignores the probability that variation in micro-environment and individual resources will affect yield.

Only a few anomalous data force him to admit, temporarily, the inadequacy of his assumptions. In the initial analysis, it was discovered that three households had achieved agricultural profits *higher* than the optimum values calculated (1963:76-77). In only these cases were the data re-examined. It was discovered that these households had achieved exceptionally high yields on one or more crops. In these cases only, it was decided to substitute individual for mean yield values. There seems to be little intellectual justification for making this adjustment for only a few cases. Because of local conditions or lack of resources, some households can expect to achieve exceptionally *low* yields for individual crops. In Desai's model, this probability is ignored. Seasonal and geographic variations in price are also ignored, in spite of the fact that any rational decision-maker is forced to estimate price in his locality at the time of sale.

His reliance on interview opinion is indicated in the following passage where he argues that the farm households studied have made inadequate use of irrigation facilities:

> The important thing which we noticed was that the farmers did not make full use of the irrigation facilities they had. In the beginning of the inquiry, the irrigable area on each farm was noted. This was done by noting down the irrigable portion of each plot on the farm. At the end of the inquiry, it was found that the area which was actually irrigated was much less than the irrigable area (1963:40).

The statement of this conclusion, on the basis of data collection techniques outlined, displays a substantial lack of caution. The assumption that the initial notations of outside interviewers are somehow "correct," and that the judgments of the farmers using the facilities are somehow "incorrect" rests on an infirm intellectual base.

Because of these drawbacks it is virtually impossible to draw trustworthy conclusions from Desai's study. The study is useful, however, in exposing some of the pitfalls risked by the student of decision-making behavior.

Farmers Viewed as Material Maximizers

Scholars who have emphasized the importance of material goals are sometimes called "formalists," apparently because they believe that formal economic theory can be applied with equal validity to explanation of behavior in both Western and non-Western societies. Many anthropologists and most economists belong to this group.

Theodore Schultz is an economist with a strong interest in problems of economic development, and with the specific role of agriculture in the development process. He is concerned with determination of optimal combinations of development techniques. A basic tenet of his approach is the material self-interest of the "traditional" farmer. His position is stated succinctly in the following hypothesis:

> There are comparatively few significant inefficiencies in the allocation of the factors of production in traditional agriculture (1963:37).

Schultz implies that traditional farmers have reached a level of efficiency in allocation of production factors through sustained effort and careful choice. This conceptualization is essential to Schultz's proposals which center on research (seeking new and more efficient production factors), provision of information (about new and improved production factors), improved distribution (of production factors), and improved rural education (particularly in the area of record keeping and evaluation of production factors) (1963:145-206).

Since 1960, economic studies which bear out the essential truth of Schultz's assertions have appeared in increasing numbers. These are studies of "supply response" in underdeveloped nations. Supply response studies have been applied to a wide variety of perennial and annual crops, including cocoa, rubber, coffee, lemons, other citrus fruits, bananas, palm oil, copra, tea, dates, olive oil, jute, rice, corn, wheat, cotton, barley, tobacco, gram, jowar, and bajar. (Bateman 1969:243; Behrman 1968:15-18). In nearly all cases, these studies attempt to discover a dynamic relationship between market prices for a particular crop, and the cropping area which is devoted to it. For example, taking an annual crop like rice, total cropped area *this* year is regarded as a function of price received last year. For perennial crops the assumed relationship is complicated by a "lag" phase of several years which tree crops require to begin bearing and reach maturity.

The models or functions which are used in an attempt to relate "supply response" to price movements are exquisitely complex. No effort will be made to discuss any of them. However, the relationships discovered are measured in terms which are amenable to explanation and discussion. Presumed relationships are expressed in terms of a "coefficient of supply elasticity." This relationship indicates the degree of correspondence between a price function and a supply function. The coefficient of supply elasticity is most often calculated as follows: E = percent that supply has increased/percent that price has risen. A positive E indicates a rational relationship between price and planting decision.

Behrman has collected data from a large number of supply response studies. All provide coefficients of elasticity for "short run" periods and a few also provide long run coefficients. The latter figures are consistently higher than the former. This is consistent with Samuelson's axiomatic statement that:

> Elasticity of supply tends to be greater in the long run, when all adjustments to the higher price have been made, than in shorter periods of time (1967:369).

This axiom reveals the impossibility of immediately reacting to price movements. But it also reveals that, given enough time, farmers do react to price movements in a rational manner.

In summarizing the results of forty-four separate studies, by more than fifteen scholars, carried out in sixteen developing nations, and concerned with eleven annual crops and four perennials, Behrman makes the following generalizations:

> First, for annual crops there is a tendency for area to be used as the dependent variable because yields are often found to be unresponsive to price changes. Second, among the annual crops there is a tendency for nonfood crops (jute, cotton, and tobacco) to be more price elastic than the food grains. Third, the perennials tend to be less elastic in the short run and require longer periods of adjustment than do the annuals, as might be expected because of longer gestation periods of the perennials. Fourth, the estimates taken as a whole lend some support to the hypothesis that price responsiveness in underdeveloped agriculture is of an order of magnitude quite comparable to that observed in the agriculture of high income areas (1968:14, 19).

For annual crops referred to by Behrman, coefficients of elasticity vary between 0 and 1.52 in the short run (reaction to price in the previous year) and 0.02 and 1.62 in the long run (allowing farmers several years to react to a "permanent" price shift). Short run coefficients are relatively high for fiber crops, 0.29-0.75 for jute, 0.41-0.59 for cotton, and relatively low for grain crops, ordinary values for rice are around 0.20 (Behrman 1968:15-18). These figures are comparable to those calculated for Western commerical farmers (Wharton 1969).

Behrman personally has calculated supply response (for 1937-1963) coefficients for four annual crops in fifty provinces in central and northeastern Thailand. The annual crops are rice, kenaf, corn, and cassava. The dependent variable (representing "supply response") is total area planted in each case for the crops. The dependent variable is primarily treated as a function of the previous year's price, but also as a function of farm population, *expected* yield, and yield variability. For example, first, and most importantly, Behrman measured the relationship between price fluctuations and fluctuation in area planted in a given crop. He also measured the relationship between fluctuations in population, expected yield, and yield variability and area planted. In each case, the degree of relationship is indicated by a coefficient of supply elasticity.

Short run coefficients of supply elasticity for rice in 50 central and northeastern provinces varied between 0.02 and 0.62 during 1940-1962. The figure for all Thailand was 0.18 for the same period. The latter figure can be interpreted as follows. If the price of rice increases 10 percent *this* year, the total area planted in rice should increase 1.8 percent *next* year. Long run coefficients of supply elasticity varied between 0.06 and 3.12 in 50 provinces. The figure for all Thailand is 0.31 (Behrman 1968:290-293). The latter figure can be interpreted as follows. If the price of rice manifests a consistent upward trend of 100 percent over a period of several years, the total area planted in rice can be expected to increase by thirty-one percent during the same period.

Behrman comes to the following conclusions:

> This study has not attempted to explore directly the complex problems which retard the long run transformation of traditional agriculture. The study has provided new and compelling evidence, however, that, in the short run, farmers in underdeveloped countries respond rationally and substantially to economic incentives. No significant evidence, in contrast, has been found for the hypothesis that institutional constraints preclude significant responses to economic incentives in underdeveloped agriculture. The burden of proof, thus, now lies with those who maintain that the supply behavior of farmers in underdeveloped agriculture cannot be understood within the framework of traditional economic analysis (1968:337).

Behrman issues an intellectual challenge to Polanyi, Dalton, Nash, Kingshill and all other scholars who assert that technologically simple farmers are motivated largely by non-material goals. If their position is correct, it should be possible to marshall quantitative data and models which support their point.

Behrman has presented massive evidence to support the conclusion that Thai farmers are no less responsive to price (and thus, no less rational) than Western, commercial farmers. The model developed in this monograph makes much the same point, but is based on different procedures, and operates at a different demographic level. Although Behrman's research falls within the range of microeconomic theory, the data handled are obviously "macro" in comparison to those gathered at Sansai. He considers his data to be relatively "disaggregated" because he makes comparisons at the provincial level (1968:2). In doing so, he compares units which include hundreds of thousands of individuals. In contrast, the model developed here applies to a village of 900 people.

Decisions under Uncertainty

Sufficient data have been presented (Chapters Five and Six) to provide a basis for the discussion of a tentative model of crop choice at Sansai. Construction of the model is dependent on use of several concepts and approaches seldom utilized in anthropology. It will be argued that Sansai farmers make decisions in a situation where Game Theory does not apply. Perhaps Bayesian statistics can be applied in the Sansai situation, but they appear to be unduly complex. Our model is based on three straightforward "procedures" and is analogous, in obvious ways, to the statisticians' T-test and Analysis of Variance.

Since the construction of the model is dependent on precise utilization of several related concepts, a series of definitions is necessary.

Decision. A choice between acts which differ in their implications, results, or payoffs.

Payoff. A "settlement or reckoning, as in retribution or reward" (*Random House Dictionary of the English Language* 1969). The notion of "payoff" is closely related to the concept of "utility."

Utility. The "capacity of a commodity or a service to satisfy some human want" (*Random House Dictionary of the English Language* 1969).

Probability. The "relative frequency with which an event occurs or is likely to occur" (*Random House Dictionary of the English Language* 1969).

Decision Under Certainty.[1] A decision made under conditions such that "each action is known to lead invariably to a specific outcome" (Luce and Raiffa 1958:13).

Decision Under Risk.[2] A decision made under conditions such that "each action leads to one of a set of possible specific outcomes, each outcome occurring with a known probability" (Luce and Raiffa 1958:13).

Decision Under Uncertainty.[3] A decision made under conditions such that an action "has as its consequence a set of possible specific outcomes, but where the probabilities of these outcomes are completely unknown or not even meaningful" (Luce and Raiffa 1958:13).

Decision Tree. A diagram which utilizes branching lines to represent a choice or series of choices.

Decision Matrix. A rectangular arrangement of payoffs or utilities such that each entry represents the conjunction of the choice of a single player and a state of Nature, or the choices of two players.

Nature. A disembodied entity, lacking both reason and will, which randomly imposes a range of alternate states on a decision-making situation.

Maximization. The goal of all behavior which involves choice under scarcity. Maximization may be directed toward payoffs which are expressed in terms of material or non-material items. In this case, of course, material goals have been emphasized.

The following three questions are directed toward the data presented in Chapters Five and Six. An answer of each question will lead to discussion of our tentative model of crop choice.

(1) *How can the decision situation which an individual farmer is faced be best represented?* Each farming household controls a plot of land utilizable in a variety of ways. Significant choices are related to rice variety, garden crop, planting date, etc. A descriptive technique is required which clearly represents this wide range of available choices. Both "decision trees" and "decision matrices" will be presented and useful insights into the importance of minor variation in altitude will be gained.

(2) *How are payoffs to be defined?* A decision is a meaningful event only so long as choices are related to variable payoffs or utilities. Data on payoffs are gathered empirically. There is a constant problem of assessing relevance of data and establishing precise definitions. The problem is choice of appropriate data from among those available. It will be argued that payoffs are defined precisely, but differently, under variable environmental conditions.

(3) *How are choices made?* Game theorists present one model of decision-making under uncertainty. In the Game Theory approach to "decisions against Nature" it is assumed that Nature is a "player" in active opposition. This assumption, if correct, enables the human "player" to reduce a situation of uncertainty to one of risk. It will be argued that the game model is unrelated to crop choice at Sansai, and that another model is required. It is more useful and appropriate to assume that Nature is *not* an active, willful player. From the viewpoint of the farmer, Nature is a bundle of probabilities, too complex to be completely known, but partially discoverable on a year-to-year basis. There are some attractions to a Bayesian approach, but it may be unnecessarily complex

for this situation. Our model is built around the notion that Sansai farmers discriminate between choices by methods analogous to those used in conventional statistics. This model is based on the explicit assumption that material self-interest is deeply embedded in human behavior.

Representation of the Decision Framework

In real life, decision-making is a diachronic, rather than synchronic process. Most outcomes result from a series of choices, rather than a single decision. At Sansai, crop choice is a process, rather than a single act. The farmer makes a choice of a garden crop or a rice variety for each of two or three seasons. He then makes additional decisions concerning timing, quantity of fertilizer, plowing method, sales method, etc. This complexity is sometimes given graphic representation in "decision trees."

Chart 18 provides a much simplified representation of annual crop choice at Sansai. It gives consideration only to choice of rice variety or garden crop, without introducing other significant variables--timing, amount of fertilizer, plowing method, etc. Even so, the number of strategies *available* to Sansai farmers quickly mounts to forty. When we consider two or three additional factors, it can be seen that Chart 18 greatly underestimates the number of available strategies. First, at every juncture at least one additional possibility exists. For example, instead of choosing a rice variety or garden crop, the farmer can choose to do nothing with the land. Occasionally, a household does so. Second, strategies can be mixed, i.e. a household controlling five *rai* need not plant all of it in a single early rice variety. It may choose to plant three *rai* in *khaaw kaew dau* and two *rai* in *khaaw kam phaaj*, or some other combination from among the wide range available.

In theory, the number of choices available rapidly approaches infinity. In fact, walled plots smaller than one-quarter *rai* are virtually always planted in one variety or one crop, so that the number of available strategies is large, rather than infinite. For each possible strategy there is an associated range of payoffs, dependent on states of Nature. Nature allows Sansai households substantial flexibility in crop choice, making the relatively narrow range of strategies actually used more meaningful. That is, the narrow range of used strategies, when compared with the wide range of possible strategies, may itself be indicative of the existence of "rationality," or something like it.

Chart 18 has been divided into three sections to clarify an important truth about crop choice at Sansai. Depending upon the altitude of the land controlled, farmers at Sansai are faced with radically different decision frameworks. On the lowest plots, flooding makes cultivation of late rice impossible; on the highest, inadequate irrigation facilities make early rice cultivation impossible. Only lands of medium altitude can be triple-cropped. Each decision framework incorporates a different set of strategies and a different range of payoffs. Although the range of payoffs for each decision framework overlaps the others to some degree, there are significant differences. Medium land is clearly superior, high land is clearly inferior, and low land is "in between" in terms of expected yield.

Decision trees provide useful insights, but students of decision theory and Game Theory find them clumsy to use. Ordinarily they summarize

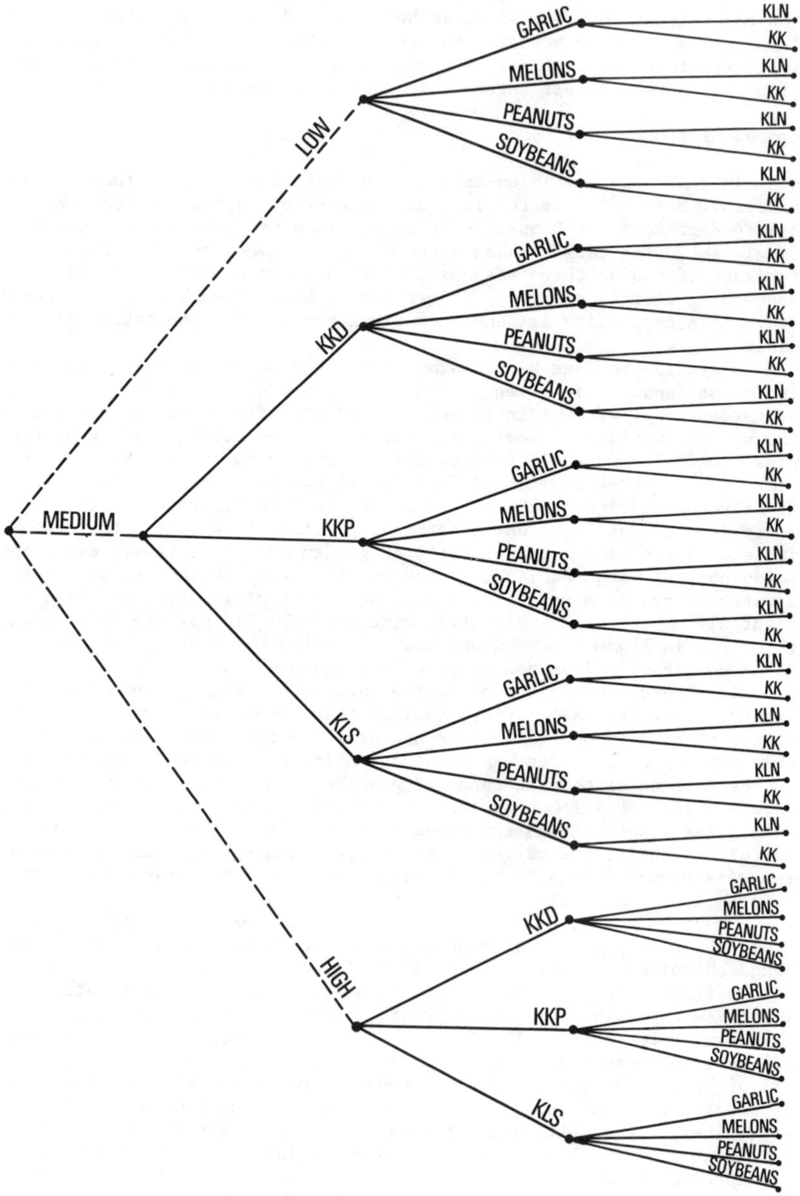

Chart 18 *Decision Tree: Crop Choice at Sansai*

decision frameworks in matrix form. (Tables 32, 33, and 34 present the problem of "crop choice" at Sansai.) The matrix form does not adequately represent all the steps in all possible strategies, but it does clearly associate strategies actually used with their appropriate payoffs. Reduction of data to matrix form is an integral part of the Game Theoretic approach to decision-making. However, it is not a necessary aspect of our model.

Table 32 *Payoff Matrix for Annual Crop Choice on Low Land at Sansai*

	Profit on Land (*baht/rai*)		
	Low*	Mean	High*
Garlic and KLN	640	975	1475
Garlic and KK	770	1160	1470
Watermelons and KLN	360	720	1135
Watermelons and KK	485	815	1185
Peanuts and KLN	375	620	865
Peanuts and KK	530	715	900
Soybeans and KLN	230	495	760
Soybeans and KK	365	585	815

* "Low" and "High" estimated payoffs were arbitrarily set at levels one standard deviation above and below the mean value. Also applies to Tables 33 and 34.

Table 33 *Payoff Matrix for Annual Crop Choice on Medium Land at Sansai*

	Profit on Land (*baht/rai*)		
	Low*	Mean	High*
KKD and garlic and KLN#	790	1270	1760
KKD and garlic and KK	930	1355	1805
KKD and melons and KLN	413	958	1483
KKD and melons and KK	553	1053	1523
KKD and peanuts and KLN	451	793	1139
KKD and peanuts and KK	582	880	1178
KKD and soybeans and KLN	296	662	1028
KKD and soybeans and KK	427	750	1071
KKP and garlic and KLN	715	1210	1725
KKP and garlic and KK	850	1305	1760
KKP and melons and KLN	362	913	1457
KKP and melons and KK	502	1002	1502
KKP and peanuts and KLN	541	900	1259
KKP and peanuts and KK	672	987	1302
KKP and soybeans and KLN	386	770	1152
KKP and soybeans and KK	517	866	1195
KLS and garlic and KLN	605	1145	1730
KLS and garlic and KK	750	1245	1770
KLS and melons and KLN	257	862	1467
KLS and melons and KK	382	957	1512
KLS and peanuts and KLN	432	850	1264
KLS and peanuts and KK	563	935	1307
KLS and soybeans and KLN	277	715	1157
KLS and soybeans and KK	408	804	1200

KKD, KK, KLS, and KKP are rice varieties.

Table 34 *Payoff Matrix for Crop Choice on High Land at Sansai*

	Profit on Land (*baht/rai*)		
	Low*	Mean	High*
KKD and garlic#	480	830	1305
KKD and watermelons#	303	578	1028
KKD and peanuts	225	410	595
KKD and soybeans	-13	279	561
KKP and garlic#	435	725	1135
KKP and watermelons#	157	487	877
KKP and peanuts	280	465	650
KKP and soybeans	96	331	506
KLS and garlic#	340	615	1025
KLS and watermelons#	57	377	752
KLS and peanuts	170	355	540
KLS and soybeans	6	225	442

KKD is more popular in Sansai than the other *naa pii* varieties because of its early maturation date. With an early harvest it is possible to plant melons or garlic early and receive an early-in-season price. This fact has been accounted for by estimating prices received for these crops at ten percent above the mean when following KKD and ten percent below the mean when following KKP or KLS.

Definition of Payoffs

In *Theory of Games and Economic Behavior*, von Neumann and Morgenstern use an axiomatic treatment to establish a definition of "utility" (1964: 20-30). They do so to widen the scope of the Theory of Games. Their goal is to establish an experimental basis for assessing utility, and then to treat the results *as if* they were quantitative data. They argue that this is mathematically justifiable. This step broadens the scope of Game Theoretic models, i.e. the "game" treatment can be extended to situations where no clearcut numerical payoffs exist.

However, the student of crop choice at Sansai is nearly overwhelmed with quantitative data. The problem of establishing appropriate units for payoffs is quite different. Farmers readily provide information on yield, selling price, area planted, labor inputs, and capital expenditure. The question arises: "For what material payoff are these farmers maximizing?" A comparison of agricultural practices in two northern Thai rice villages is incorporated to provide a tentative answer to this question. It will be argued that *some* farmers, in lightly populated areas, maximize for "productivity," while others, in areas of dense population, maximize "yield." These two terms are defined as follows.

Productivity. Farmers who seek to maximize "return on labor" are orienting their behavior toward "productivity." In northern Thailand we might choose to evaluate behavior oriented toward this goal in terms of *baht* per man-day.

Yield. Farmers who seek to maximize "return on land" are orienting their behavior toward "yield." In northern Thailand we might choose to evaluate behavior oriented toward this goal in terms of *baht* per *rai*. Under particular circumstances. a crop, variety, or technique may score relatively high on one scale, and relatively low on the other.

A comparison of agricultural practices at Ban Ping and Sansai is appropriate for our purposes. When studied by Moerman in 1960-61, Ban Ping's 120 households farmed 2470 *rai* of field land, with a population density of roughly 150 persons per km^2. This sizable difference in land:population ratio is reflected by substantial variation in agricultural practice. Table 35 permits the following five generalizations.

First, Sansai's fields are much less varied than fields at Ban Ping which Moerman refers to as distinct "micro-environments." Some fields at the latter site are recently cleared and quite far from the village.

Second, Ban Ping's agricultural schedule revolves around a single crop--glutinous rice. Through careful manipulation of variety, planting date, and diverse "micro-environments," it is possible to extend the rice cropping season to eight or nine months. The agricultural schedule at Sansai revolves around double- and triple-cropping. In spite of this, there are agricultural "lulls" in Sansai in May-June and September-October.

Third, at Ban Ping, rice cultivation is largely restricted to glutinous varieties, although it is not clear whether they are photosensitive or not. At Sansai, only the late rice crop is planted in glutinous types.

Fourth, at Ban Ping, tractors are used at the newer, more distant fields. In fact, it was the introduction of tractors into this area which made their clearing and cultivation possible. At Sansai, tractors are rented only when time is short. Occasional use of tractors for plowing increases the feasibility of double- and triple-cropping.

Fifth, large areas at Ban Ping are sown, and not transplanted. This seems to be part of a general intent to minimize labor inputs at distant fields.

Table 35 *Agricultural Decision-Making at Ban Ping and Sansai*

	Ban Ping	Sansai
Population density[4]	150/Km^2	670/Km^2
Crop land per household	19.8 *rai*	3.3 *rai*
Crop land per person	3.95 *rai*	0.69 *rai*
Major problem	Mobilization of labor	Gaining access to land
Major agricultural goal	Maximization of labor productivity	Maximization of yield
Variation in field type	Fields at six locations, each watered or irrigated in a different way, the nearest about 10 minutes from the village, the farthest several hours.	"High," "medium," and "low" fields, all within minutes of the village, all watered by a single, government-controlled irrigation scheme.
Agricultural schedule	Diversity of field types and varieties with different maturation periods enable farmers to keep busy from late April to December. Sales and transport of rice, repair of tools, etc. occupy the other months.	Double- and triple-cropping keep many households busy during the year. A "garden" crop is raised in addition to one or two of rice. Are lulls in May-June and September-October.

Table 35 (Continued)

Major Cropping Decisions	Ban Ping	Sansai
Choice of variety	Moerman discusses distinction between glutinous and non-glutinous varieties, and those varieties with short, medium, and long growing seasons, but makes no effort to evaluate specific varieties. Makes no mention of photosensitive/non-photosensitive distinction.	Glutinous varieties are used for late rice and non-glutinous for early rice. Three specific varieties for the former, two for the latter. Also choice of four garden crops. Late rice varieties are photosensitive, others are not.
Plowing	Tractors are used on some fields, but not on others. Preferred on fields which are distant, where labor is difficult to mobilize, and the soil is plowed dry.	Tractors are used occasionally before late rice, often before watermelons and garlic, and seldom before peanuts, soybeans, or early rice. Preferred when time is short.
Planting	Some fields are transplanted, others simply sown.	All rice is transplanted. Even melons are transplanted.
Use of fertilizer	Commercial fertilizer is sometimes used on seedlings in nursery. Stubble burned.	Commercial fertilizer is used for garlic and melons, not for the rice crops, peanuts or soybeans.

Table 35 clearly demonstrates the existence of distinct agricultural strategies at Ban Ping and Sansai. From this Table we gain the impression that strategic behavior varies, in part at least, because it is directed toward different goals. A discussion of land clearance, purchase, inheritance, and rental will throw greater light on this matter. Further insights will be gained through a discussion of patterns of labor expenditure and mobilization.

The Ban Ping area was pioneered about 1860 by ancestors of the present residents. While there is evidence that they were not the first settlers, the area was virtually unsettled when they arrived.

> The founders of Ban Ping left Chiengmuan for the plentiful land of Chiengkham. The remains of an old temple indicate that there probably had been a village near Ban Ping before the present settlement. The oldest villagers report that their parents told them that some land had already been cleared on the Great Field when the Lue first arrived. There was, they say, already an irrigation system which merely needed improving and enlarging, but most of the Great Field was unclaimed land. The jungle pressed close about the village, and tigers and monkeys abounded (Moerman 1968:91).

In time, all land at the Great or Home Irrigated field was claimed and developed. Before this happened the villagers allowed for "uncultivated roadways wide enough for carts" (Moerman 1968:91). After the Home Irrigated field was fully claimed and cleared, farmers moved on to clear plots at the Semi-Irrigated and Rainfall fields. Plots are still being cleared in these locations. Ban Ping residents first attempted to clear land at the Home Flood field about 1925. They attributed the failure of this effort to a high concentration of spirits. Since 1949, the area has been cleared and cultivated successfully. Since settlement in 1860, Ban Ping residents have had rather consistent access to virgin land. As a result, one century after the village was founded, the per capita land supply is 3.95 *rai* (Moerman 1968:91-92).

When the grandparents of Sansai's eldest male resident moved to Sansai around 1860 the village had only seven houses. The present populations of Sansai (905) and Ban Ping (approximately 600) have grown from similar bases to their present levels in about 100 years. However, Ban Ping residents have had constant access to virgin land that Sansai residents have lacked for two generations. The village's eldest male resident, Uuj Myyn, raised tobacco on cleared "forest" (*paa*) land around 1910, on or near the present site of Longphumoon, a village located less than one kilometer from Sansai. Younger men in their 50's and 60's have devoted their lives to working purchased, inherited, or rented land, indicating that virgin land has long been a part of Sansai's past. The nearest unclaimed, forested areas lie perhaps twenty kilometers away and dozens, if not hundreds, of villages lie closer to the forests than Sansai.

At present, and for the two past generations, Sansai residents have been able to increase their land base only through intensification. *Caaw* Mahawoong introduced double-cropping of rice on some Sansai fields in the late 19th century, apparently as a measure to increase the revenue base. Informants report that garlic has been raised as a second or third crop since about 1910. Sansai residents have intensified land use through double- and triple-cropping for two or three generations.

The relative scarcity of land at Sansai can be demonstrated in other ways. While a great deal of land is acquired through inheritance at Ban Ping, the preponderant method until 1960 was clearing and claiming. More than eighty-five percent of the 1375 *rai* acquired at Tunglor was cleared and claimed between 1955 and 1960. In addition, large quantities of land are bought and sold at Ban Ping. For example, most of the twenty-four plots owned at the Distant Irrigated field were purchased from Yuan owners (Moerman 1968:91-96). At Sansai land prices are beyond the ability of ordinary farmers to pay. The little land purchased during the 1960's was bought by shopkeepers, middlemen, and urban residents, not by farmers.

At Sansai, inheritance of land sometimes leads to disputes among siblings which are ordinarily settled by the village headman (*kamnan*). In some cases the disputed shares are as small as one or two *rai*. In contrast, at Ban Ping

> It is unnecessary to consider complexities of inheritance by minors or the distribution of plots of uneven quality. The supply of land is sufficient, so that such issues are not contested (Moerman 1968:94).

At Sansai, these issues, and others, provide a basis for inheritance disputes. Siblings become bitter enemies over disputes involving very small plots. They accuse each other of lack of responsibility toward their dead parents and similar wrongs. Such disputes are a clear indication of land shortage.

Land is seldom rented at Ban Ping. The proportion of rented to owned plots worked at various fields in 1960 is given below (Moerman 1968:100-103).

Field	Total Plots	Number Rented	Percent Rented
Distant Flood (*Tunglor*)	103	1	1.0
Distant Irrigated	29	1	3.4
Home Irrigated and Semi-Irrigated	64	14	21.9
Home Rainfall	31	0(?)	0
Home Flood	6	0(?)	0
Total	233	16	

Moerman indicates that no renter is landless, and that most rentals are arrangements of convenience. Many landlords are also tenants. They attempt to consolidate their holdings at one field, or to distribute labor requirements more evenly. Occasionally a farmer rents to "help" the landlord (Moerman 1968:100-103). By contrast, landless renters form a major sector of Sansai's population. Exactly fifty percent of Sansai's 186 households own no land or house land only. Many of these rent land during one or more agricultural seasons.

Rents are generally lower at Ban Ping than at Sansai. Kinsmen pay one-third of their total production to the landlord, while fellow villagers pay two-fifths. Only outsiders are expected to pay one-half of the total production (Moerman 1968:111-112). At Sansai a pattern of half-shares has been established for several decades. Rental arrangements, as well as the large proportion of renters, emphasize the scarcity of land at Sansai and its high value.

If land resources at Ban Ping are ample, and those at Sansai are severely limited, we must make the opposite statements about labor resources. In several ways Ban Ping residents indicate that their labor supply is limited, while the opposite situation holds at Sansai. Whether we consider total labor inputs, or expenditures on specific tasks, indications exist that suggest that the labor supply is "tighter" at Ban Ping. Labor inputs at Ban Ping and Sansai are given below (Moerman 1968: 202).

Rice Crop	Labor Input (Man-days)
Animal-plowed, transplanted rice, Ban Ping	15.8
Early rice, Sansai	24.4
Late rice, Sansai	22.3

Available data suggest that labor inputs at Sansai are 40 to 55 percent higher. Because yields at Sansai and Ban Ping are in the same general range, productivity at Ban Ping is substantially higher. Labor inputs at Ban Ping are low in large part because no weeding is done (Moerman 1968:162). The average 6.14 man-days per *rai* which Sansai residents devote to weeding the early rice crop and 5.11 man-days for the late rice, account for most of the difference in labor expenditures between the two villages. It is likely that Sansai residents could increase their productivity by omitting all weeding operations but at some cost to yield. That they do not do so provides clear indication of their position; they have no virgin lands and no other fields to apply their labor. Given their position, Sansai residents are rational in choosing a 40 or 55 percent increment in labor costs. If it provides a mere 10 to 20 percent increment in yield, the effort is well justified. At Ban Ping, where land is plentiful, such reasoning is less obviously rational.

At Sansai, where numerous agricultural workers usually are available, labor mobilization is a casual process. Farmers who require extra workers (to transplant or harvest rice or to plant garlic or watermelons) ordinarily hire workers a day or two in advance, and nearly always on a day-to-day basis. Payment is in cash at the rate of five to seven *baht* per day. Only two households hire agricultural workers for more than several days at a time. In both cases the farmer fulfills social responsibilities by hiring one or more permanent workers. One young man who lives with his elderly parents is hired by his elder brother on a seasonal basis. At the end of each rice cropping season he receives payment in paddy. The elder brother views the arrangement as fulfillment of an obligation to his family of origin. The headman hires seven workers for about seven months each year. However, these laborers work at Doi Saket, far from the village, near the edge of the Chiang Mai plain. The headman owns land there, and the workers do no work within the village. This arrangement allows the headman to fulfill more completely his role as a member of the traditional *caaw* lineage of northern Thailand. In other words, the employer/employee relationship takes on overtones of a traditional patron/client relationship.

At Ban Ping, both residents and migrants from Laos are hired as seasonal laborers. In 1960 some thirteen households hired workers for the tasks of irrigation, seeding, plowing/harrowing, and transplanting. In twelve of these cases the "tillage season" laborer was a near or distant relative of the household head. Workers often describe their employment in terms of "helping" a father, father-in-law, or uncle (Moerman 1968: 118-119). At harvest time Ban Ping residents depend on migrants for part of their labor supply. Moerman reports that:

> In 1960, as in every year for as long as villagers can remember, a few young Lue men from Myang Khawp in the contiguous part of Laos came to Chiengkham seeking farm work (1968:137).

At an earlier time, many of these young men were hired for an entire agricultural year. By 1960 a new pattern had evolved. Young men from Myang Khawp were hired only for the harvest season and most of them worked at distant fields. Local labor is easily mobilized at fields close to the village, but "selling rice" at the Distant Irrigated and Distant Flood fields is harvested almost exclusively by household members and "harvest season" laborers (Moerman 1968:137).

Through utilization of seasonal labor contracts and hiring of migrants, Ban Ping residents demonstrate that labor, relative to land, is in short supply. By relying on daily laborers and occasionally exporting harvest labor, Sansai residents demonstrate the opposite relationship between land and labor.

In both villages rice is the major crop and source of food. However, within the category of "villages which produce wet rice with simple tools" the two lie near opposite extremes. As late as 1960, Ban Ping remained a pioneer village with much land awaiting development or additional improvement. Sansai has reached the status of an emigrant village. Land resources are extremely limited and the prices are extremely high. No new paddy fields have been developed in two generations. By controlling land at several locations, Ban Ping residents are able to manipulate agricultural schedules, and spread a limited supply of labor over extensive land holdings. Through double- and triple-cropping, Sansai residents are able to obtain a maximum product from a limited supply of land.

Construction of an adequate model in the next section is dependent on careful definition of the goals of agricultural decision-making. These comparative data place agricultural goals at Sansai in perspective. Relative to residents of Ban Ping, Sansai farmers place greater emphasis on maximization of yield. This being the case, alternate agricultural strategies at Sansai are best measured in terms of "buckets per *rai*" rather than "buckets per man-day."

Decision Models: Game Theory and "Decisions Against Nature"

A number of solutions have been proposed for decision problems under uncertainty, in which choices must be made in the face of payoffs of unknown probability. The most widely known approach is Game Theory.

> In a game the uncertainty is due entirely to the unknown decisions of other players, and, in the model, *the degree of uncertainty is reduced through the assumption that each player knows the desires of the other players and the assumption that they will each take whatever actions appear to gain their ends* (Luce and Raiffa 1957:275, italics added).

Game Theory is applicable only to those situations in which all parties are assumed to be self-interested maximizers. Situations in which a single decision-maker must develop strategies in relationship to a disinterested Nature or Society are not amenable to this kind of analysis.[5] Some other approach must be found. A variety of approaches have been suggested (e.g. by Savage, Chernoff, and Hurwicz) for uncertain situations in which no information is available at all (Luce and Raiffa 1957:280-284). But situations of partial information invite still other solutions. The model developed in this Chapter is one such solution.

In cases where a choice is made under uncertainty, it is useful to assume that the decision-maker has mentally reduced uncertainty to risk. That is, the individual proceeds *as if* with full understanding of the number and probability of possible outcomes. This is ordinarily accomplished by treating partial knowledge *as if* it were complete. The conditions under which farmers usually make decisions exemplify this situation. Crucial choices must be made in advance of the cropping season on the basis of best possible estimates of climatic conditions, price levels, yields, etc. In this situation, uncertainty can never be reduced to risk by acquiring complete knowledge. But additional knowledge does result from each year's experience, and each year the farmer must proceed *as if* relevant information is complete. Situations of this type--in which knowledge is incomplete and there is no basis for assuming the existence of a self-interested, opposing player--are here labeled as "Decisions Against Nature." Our model exemplifies one possible approach to decisions of this type.

The Model: "Specific" Form

Our model leans heavily on several concepts basic to conventional statistical analysis, e.g. *sampling*, *mean*, *variability*, and *normal distribution*. It is assumed, however, that the Sansai farmer is motivated by material self-interest rather than the "scientific sobriety" which is thought to underly statistical analysis. In essence, this means that the farmer is faced with a situation in which *some* choice must be made, while the social scientist often is allowed to state that "No choice is obviously correct." Whereas the statistician or statistically-oriented social scientist demands a high level of likelihood (e.g. significance levels of .05 or .01) before he accepts that $A \neq B$, the farmer must pay careful attention to *any* indication that $A > B$ or $A < B$.

Table 36 presents data on the rice crop under discussion. Available data indicate rapid adoption of *khaaw kaseet*. This rice variety was unknown at Sansai in 1968, yet it was the most popular early variety by 1971. An adequate model should predict this rapid acceptance on the basis of available yield figures. At the same time, the model should predict that some households will change, while others will not. The possibility of changing from *khaaw kaseet* to *khaaw lyang nauj* should not be excluded; "perverse" behavior of this sort occurs in the real world. In summary, tne model should account for the fact that decision-making is manifested by "trends" rather than group movements from one strategy to another.

The conditions set out above are satisfied by assuming that farmers, like social scientists, learn about the world through sampling procedures. The crop planted each year constitutes a single experiment. The farmer learns the outcome of additional crop experiments by talking with friends and neighbors. Our model incorporates three "procedures" into a decision framework. Annual repetition of these procedures provides a dynamic (stochastic or iterative) model of crop choice which sometimes predicts movement toward an equilibrium of choice, sometimes predicts sudden change.

Table 36 *Early Rice Varieties Selected in Recent Years**

Year	Households Planting *Khaaw Lyang Nauj*	Percent	Households Planting *Khaaw Kaseet*	Percent
1968	61 (?)	100.0	0	0.0
1969	60 (?)	98.4	1	1.6
1970	55	89.1	6	10.9
1971	11 (?)	18.0	50 (?)	82.0

* Figures for *khaaw lyang nauj* in 1968, 1969, and 1971 are estimates. The first two figures are based on the assumption that equal numbers of farmers raised the crop in 1968, 1969, and 1970. Figures for 1971 are based on plans at the end of the 1970 season. By 1976 *all* Sansai farmers were planting *khaaw kaseet*.

The three successive procedures are: sampling; computation; and choice. Each procedure is discussed below.

Sampling

The model is based on the assumption that Sansai farmers gather data relevant to crop choice in a manner analogous to that used by social scientists, i.e. they draw a random sample. It is assumed that the farmer talks to his peers about their yields on a nearly random basis. For purposes of discussion, it is assumed that each farmer draws a random sample from among twenty of his peers.[6] For early rice, 1970, this implies a sample of one-third of all households raising the crop.

In 1970, fifty-five village households raised *khaaw lyang nauj*, while six grew *khaaw kaseet*. Members of the former group are assumed to draw from the sample $X_1, X_2, \ldots X_{54}, Y_1, Y_2, \ldots Y_6$, while members of the latter draw from $X_1, X_2, \ldots X_{55}, Y_1, Y_2, \ldots Y_5$. Households raising *khaaw lyang nauj* can expect to gather data from 14-20 households raising the same variety, and 0-6 raising *khaaw kaseet*. "Typically," 18 households raising the former, and two raising the latter will be interviewed. Households raising *khaaw kaseet* can expect to gather data from 15-20 households raising *khaaw lyang nauj*, and 0-5 raising *khaaw kaseet*. "Typically," they will speak to 18.33 households raising the former, and 1.67 raising the latter.

Computation

Before a final choice is made, the sample data for each variety must be computed or combined in some manner suitable for comparison. It is assumed that the farmer will have simultaneous interest in the *typical* performance of a variety, and in the *specific* performance achieved on his plot. His computational needs go beyond mere calculation of sample means. Some procedure must be developed for assigning relative "weights" or importance to data concerned with typical and specific performance.

For *khaaw lyang nauj* the farmer has data on his neighbors' yields for the year in question and several past years. It is reasonable to assume agreement as to typical performance of the established variety. In other words, after ten or twenty years the sample means of all local farmers will have converged; there will be substantial agreement concerning expected mean yield and variability. At the same time, each farmer has knowledge of *his* yield under conditions prevailing on his plot. We can expect farmers achieving yields far above, far below, and equal to the mean yield to evaluate the established variety differently.

While all farmers expect a typical yield of fifty buckets per *rai* from *khaaw lyang nauj*, those who receive yields of twenty, fifty, and ninety will evaluate this information differently. We expect them to use the same computation procedure, but to arrive at different evaluations. The computation procedure suggested is summarized in the following two formulae:

No. 1

$$V_e = \frac{2(48.8) + X_p}{3}$$

Where V_e = computed value for the established variety (*khaaw lyang nauj*)

X_p = "personal" yield value achieved by the farmer

48.8 = best available estimate of "typical" performance, based on 1970 early rice survey

No. 2
$$V_n = \frac{Y_i + 5Y_p}{n + 5}$$

Where V_n = computed value for the new variety (*khaaw kaseet*)

Y_p = "personal yield value achieved by the farmer

Y_i = sample value

Each formula incorporates a "weighting" procedure.

The computed value for the established rice variety confers twice as much significance on typical as on specific performance. Farmers have had less opportunity to gather data on the performance of *khaaw kaseet*. The computation procedure used is analogous to that for *khaaw lyang nauj*, but not identical. Too little time has passed for establishment of typical performance values. Formula No. 2 also incorporates a weighting procedure. It confers five times greater value on the farmer's personal experience than on any piece of collected data.

Choice

The choice procedure applied in this analysis is simple-minded. It is assumed that at the end of each agricultural season the farmer compares V_e and V_n, and chooses the variety represented by the higher value for the following year. Thus:

(1) if $V_e \geq V_n$, choose V_e

(2) if $V_e < V_n$, choose V_n

Very simple, these decision rules order the farmer to "Choose the variety with the higher computed value."

The model is summarized in the following outline:

Sampling. Draw a random sample (N = 20) from among $X_1, X_2, \ldots X_{54}, Y_1, Y_2, \ldots Y_6$ or $X_1, X_2, \ldots X_{55}, Y_1, Y_2, \ldots Y_5$.

Computation. Compute values for the established and new variety according to the following formulae:

(1) $$V_e = \frac{2\ (48.8) + X_p}{3}$$

$$(2) \quad V_n = \frac{Y + 5Y_p}{n + 5}$$

Choice. Choose the superior variety according to the following decision rules:

(1) if $V_e \geq V_n$, choose V_e

(2) if $V_e < V_n$, choose V_n

The sampling procedure introduces variability into the model; a range of possible outcomes is predicted rather than a single outcome. The model is tested by comparing its predicted range against a single, "real world" outcome. If the latter falls within the former, we have at least weak confirmation of the model's adequacy. Unfortunately, there is no statistical procedure which specifically evaluates a test of this type.

In this particular case, 1970 yield data are utilized to predict varietal selection in 1971. A range of predictions is generated by the process called "simulation." The specific simulation method used here involved shuffling of data cards, and twenty repetitions of each simulation.[7] The resulting predictions fall into a range or more- and less-probable outcomes, easily described in terms of mean and standard deviation.

Table 37 represents a simulated version of the model in which twenty repetitions were used. The simulation used in Table 37 predicts as "most likely" that 81.5 percent (49.7) of the households will opt for *khaaw kaseet* in 1971, and 18.5 percent (11.3) for *khaaw lyang nauj*. Based on the simulation, it can be predicted that outcomes will lie between 46.1 and 53.3 households 90 percent of the time.

The data are graphically represented in Chart 19. Like Table 37, Chart 19 demonstrates that predicted results fall within a relatively narrow range. It is possible to assign probabilities to various portions of the range. For instance, the probability that the model will predict a result between 45 and 48 households is roughly 0.20, i.e. it is the portion of the range which falls between 0.78 and 2.14 standard deviations below the mean.

Actual choices for Sansai's 1971 early rice crop are not known, but an estimate can be made with considerable confidence. Of the households surveyed in 1970, some thirty-five respondents reported their plans for the following year. Of these, only two planned to plant *khaaw lyang nauj*; the rest had chosen *khaaw kaseet*. Under the circumstances, any estimate must be considered crude, but *khaaw kaseet* is clearly the choice of the majority. Even after taking into account that some twenty households did not reveal their plans, a tentative figure of 70 to 90 percent seems justified. As outlined above, the decision model admirably fits this estimate. It predicts that 70.7-92.3 percent of Sansai households raising early in 1971 will opt for the new variety.

Because yield data are available only for 1970, there is no hope of achieving a closer fit. However, the fit achieved should not be discounted. It utilizes data from a single year, assumes a high level of rationality on the part of Sansai farmers, and correctly predicts a major turnabout in agricultural strategy. In other words, available data indicate that households raising *khaaw kaseet* constituted a small minority in

Table 37 *The Model in "Specific" Form: Simulation Repeated Twenty Times*

	Number of Households Planting, 1970	(%)	Predicted Number of Households Planting, 1971	(%)	Standard Deviation	90 Percent Confidence Interval	90 Percent Confidence Interval
Khaaw Kaseet	6	9.8	49.7	81.5	2.2	46.1-53.3	44.0-55.4
Khaaw Lyang Nauj	55	90.2	11.3	18.5	2.2	7.7-14.9	5.6-17.0

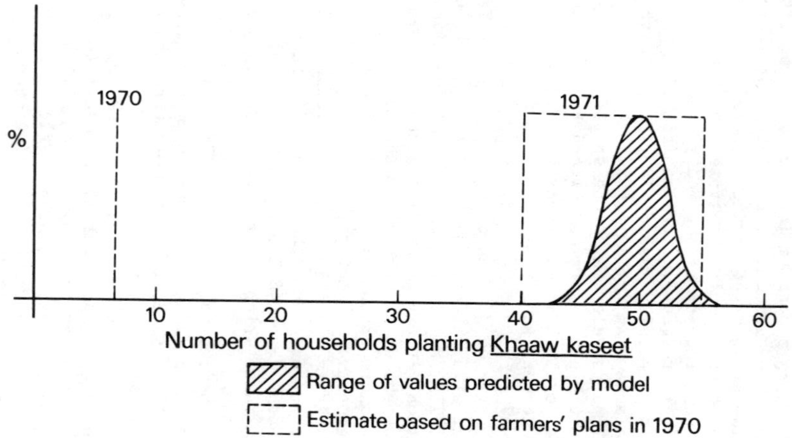

Chart 19 *The Model in "Specific Form: Simulation Repeated Twenty Times*

1970 (six of 61), and a substantial majority (perhaps 40-55 of 61) in 1971. The model clearly predicts this major shift.

The Model: General Form

As previously outlined, the model incorporates highly specific notions concerning sampling, computation, and choice procedures. In this specific form, the model generates a relatively narrow range of possible outcomes. By altering these procedures in several ways, it is possible to generate a series of analogous or related models. Sample size can be altered, computation procedures can be designed which place more or less emphasis on personal experience, and an element of "conservatism" can be introduced into the choice procedure. Ultimately, a general model can be developed, one which is applicable in a wide range of subtly varying situations.

For example, our model makes a number of strong assumptions--that all farmers draw samples of the same size, that computation procedures follow a single pattern, and that *no* farmer can be characterized by innate conservatism. An adequate general model would allow each of the procedures to vary within specified limits, and would reveal any given precedure as a particular example drawn from a range of possible types. Generality opens the possibility of wider application. With suitable modifications the model might be applied to other cropping seasons at Sansai and to agricultural decision-making in other rural areas.

Each of the three procedures (sample size, computation, and choice) can be varied and generalized. The result of generalizing procedures is, in some cases, a wider range of predicted outcomes. However, within the limits proposed, changes in variability and mean values are not large.

Sample Size

Our model is based on the assumption that each farmer discusses yield with exactly twenty friends and neighbors. Obviously this assumption does not coincide with reality, although it may be an adequate approximation. Adequate observation on this matter may be impossible, but it is also likely to be unnecessary. Chart 20 demonstrates the effect of substituting various sample sizes into the model. Increasing sample size of the model has the effect of increasing the predicted number of Sansai households raising *khaaw kaseet* in 1971.[8] Decreasing the sample size has the opposite effect. In any case, the variability of predicted outcomes is little affected.

A generalizing assumption which perhaps represents reality more accurately than any specific value is one which incorporates the notion of "random samples of randomly varying size." For example, sample size might be allowed to vary between five and fifty according to some chance event. This approach at least takes into account that sample sizes are variable. But the effect of this more general assumption on the model is minor. In essence, it approximates a specific model in which the sample size is 27.5. The range of estimates for Sansai households raising *khaaw kaseet* in 1971 is raised slightly. Variability is little affected.

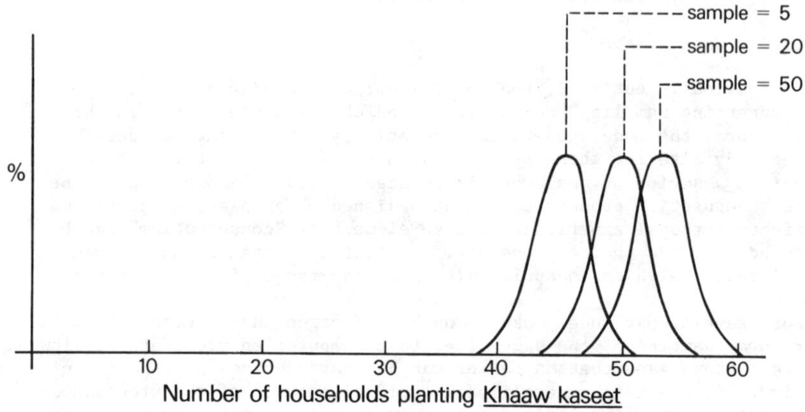

Chart 20 *Effect on the Decision Model of Using Various Sample Sizes*

Computation

Computation procedures can be varied in much the same way that sampling can. A wide range of computation procedures are possible. The procedure presented in our model assigns relative weights of 1:2 to personal and typical yield values. Other ratios--1:3, 1:5, 2:1, etc.--are possible, and they may be applicable in specific situations.[9] The significant fact is that the weighting procedure is subject to distinct upper and lower limits. Typical or average yield figures can be ignored, giving a ratio of 1:0. Alternatively, personal yield can be ignored, providing relative weights of 0:1.

Chart 21 demonstrates the implications of choosing either extreme procedure, and various ratios--2:1, 1:1, 1:2--which fall between. Each alteration of computation procedure has the effect of raising or lowering the range of predicted outcomes, but has no effect on variability. Where personal data are completely ignored (0:1), the predicted range is 44.5-57.9 households.[10] Where typical yield data are completely ignored (1:0), the predicted range is 37.5-52.3 households. Where the two types of data are given equal weight (1:1), the predicted range is 42.6-55.4. In all cases, the proportion of Sansai households predicted to choose the new variety is well above one-half.

Choice

The notion of conservatism can be introduced into the choice procedure. Rules (1) and (2) were applied in our model. Rules (1') and

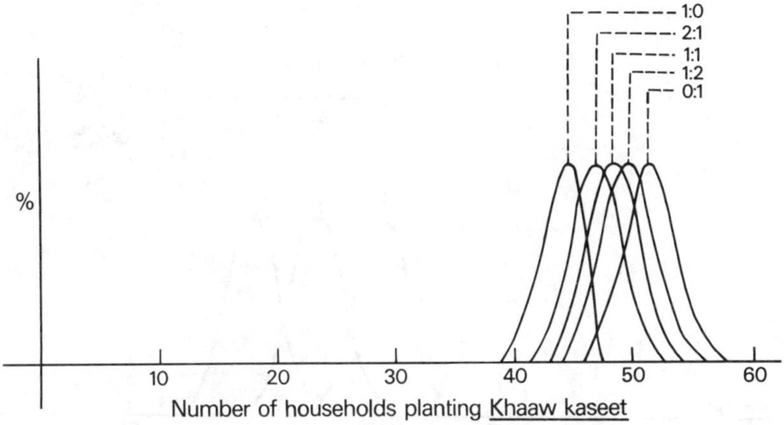

Chart 21 *Effect on the Decision Model of Using Various Computation Procedures*

(2') introduce an element of conservatism. Rules (1") and (2") are more conservative:

(1) if $V_n > V_e$, choose V_n

(1') if $0.95\ V_n > V_e$, choose V_n

(1") if $0.90\ V_n > V_e$, choose V_n

(2) if $V_n \leq V_e$, choose V_e

(2') if $0.95\ V_n \leq V_e$, choose V_e

(2") if $0.90\ V_n \leq V_e$, choose V_e

The model can be made more "conservative" by substituting rules (1') and (2') or (1") and (2") for (1) and (2).

Chart 22 demonstrates the effect of various levels of conservatism on the model. Introducing conservatism into the choice procedure has the effect of lowering the number of households predicted to choose *khaaw lyang nauj* in 1971. Nevertheless, even when the "discount" level is raised to ten percent, the proportion of households predicted to choose the new variety is well above one-half.

The choice procedure can be generalized by assuming that farmers randomly discount gathered data on the new variety by zero to ten percent. That is, some farmers are very conservative, and demand that sample data on the new variety be ten percent better before they will make a change.

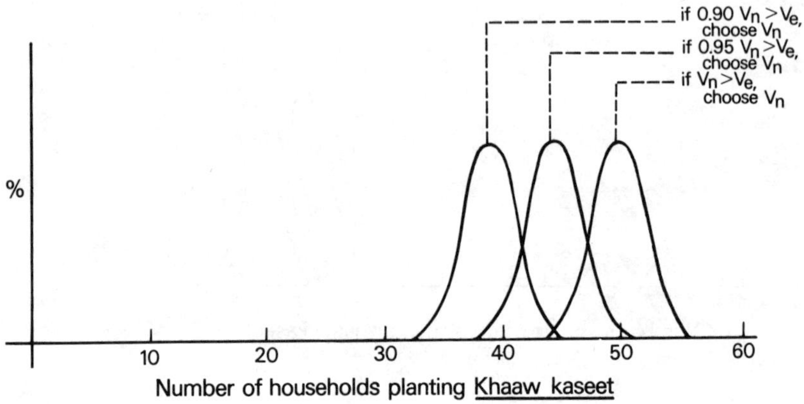

Chart 22 *Effect on the Decision Model of Using Various Choice Procedures*

Others are not conservative, and choose the variety which is supported by superior sample data. Others will fall in between. If this random, generalized procedure is used, the predicted range of outcomes will very nearly coincide with that for the "five percent discount" group above.

Charts 19, 20, and 21 demonstrate the effect of introducing generalizing assumptions, one procedure at a time, into the model. Even when upper and lower limits of variation are set relatively far apart, the effect of procedural alteration is relatively small. The model retains a strong tendency to predict that a *majority* of Sansai farmers will choose *khaaw kaseet* in 1971. Future research is necessary for precise definition of formulae and constant values, yet a model which assumes the existence of material self-interest among Sansai farmers has a sound base.

The model assumes that economic rationality exists, and emphasizes the importance of quantitative data collection. If the farmer is truly rational about crop choice, he requires quantitative data to work with as does the social scientist who chooses to analyze the farmers' "decisions against Nature." Although methods for precise testing of the model are unavailable, it is clear that reality and the model's range of predictions roughly coincide. The model clearly predicts the rapid and sizable shift in rice variety selected which occurred at Sansai in 1970-71.

Chapter Eight

IMPLICATIONS OF THE MODEL

Over generations culture and population have evolved in tandem within the Sansai ecosystem. A careful examination of Sansai culture suggests that substantial sectors of local behavior are adaptive, i.e. under present circumstances they play an important role in population maintenance. If a small number of activities in production, processing, storage, preparation, or distribution of food items were to be abruptly altered or terminated, Sansai's population would be subjected to strong, negative pressures. It would decrease through emigration, starvation, or strife or be maintained only through substantial, on-going inputs of outside aid.

The following behavioral features probably can be considered adaptive (population maintaining) in the Sansai context. An intricately adjusted agricultural schedule which makes double- and triple-cropping possible. Irrigation societies which are responsible for construction, maintenance, water distribution, prevention and control of floods, and adjudication of water related disputes. A transport system that relies heavily on small trucks and motorcycles and is remarkable efficient in the movement of farm products, capital inputs, household items, retail products, and people.

Other adaptive features of this village include a series of marketing mechanisms for locally produced crops that assure relatively rapid and smooth flow of produce into regional, national, and international markets. An intensive use of land resources for non-crop food production. Techniques of kitchen gardening, gathering, and fishing that are applied to house compounds, irrigation ditches, fallow fields, cremation grounds, the school yard, and road side areas. Items derived from these activities provide important supplements for rice-based human diets, important sources of feed for domestic animals, a cuisine in which small amounts of meat, fish, vegetables, fruit, and condiments are painstakingly introduced to make feasible a diet highly dependent on glutinous rice.

Depending upon the particular interests of the reader, the preceding list can be extended almost indefinitely. No reference has been made to the adaptive nature of local patterns of political organization, land tenure, settlement, kinship, marriage, inheritance, extension of credit, retail marketing, medical treatment, population control, or ceremonial practice. Presumably all these behavioral sectors have specific aspects which are potentially population maintaining.

In recent years, statements of the probable adaptive nature of specific aspects of contemporary cultures have become common. Harris has re-evaluated available data on India's doctrine of *ahimsa* which prohibits slaughter of cattle. He finds that Indian cattle provide an important source of traction (to grain farmers), manure (for fertilizer and fuel), hides (supporting the world's largest leather industry), milk (for the upper castes), and beef (for the lower castes and harijans). He marshalls a substantial body of data suggesting that maintenance of the Indian cattle population at present levels probably is essential to maintenance of the human population at present levels (1966:51-66). He concludes that:

> The probability that India's cattle complex is a positive-functioned part of a naturally selected *ecosystem* is at least as good as that it is a negative-functioned expression of an irrational ideology (1966:59).

Rappaport in *Pigs for the Ancestors* adopts a similar stance. His generalizations, applying to a much smaller group, are based on fieldwork rather than library research. But Rappaport concludes that certain aspects of Tsembaga Maring religion are no less adaptive than the Hindu doctrine of *ahimsa*. He summarizes his findings in this way:

> It has been argued that the regulatory function of ritual among the Tsembaga and other Maring helps to maintain an undegraded environment, limits fighting to frequencies that do not endanger the existence of the local population, adjusts man-land ratios, facilitates trade, distributes local surpluses of pork throughout the regional population, and assures people of high-quality protein when they most need it (1968:224).

Harris and Rappaport argue convincingly that a "fit" exists between cultural practice, environment, and population level. Data on agricultural schedule and cuisine at Sansai support the same conclusion. However, it is important to notice what statements of this type do *not* accomplish. They provide limited insights into adaptive *process*. They tell one a great deal about maintenance of systems, but little about their evolution. In reference to the "positive-functioned" doctrine of *ahimsa*, Harris limits himself to the statement that:

> It probably represents the outcome of intense Darwinian pressures acting upon bovine population, cultigens, wild flora and fauna, and social structure and ideology (1966:52).

We learn nothing of the origin or evolution of the *ahimsa* doctrine. Rappaport makes the same point about his own work:

> The systematic role of ritual in the ecology of the Tsembaga has been the focus of this study and *I have offered no suggestions concerning the origin of these rituals or any of the other phenomena described here* (1968:230, italics added).

He emphasizes the homeostatic aspects of the system, while paying little attention to its evolutionary component.

Neither author can be criticized for following his interests; both have made significant contributions to an ecological/materialist interpretation of human behavior. Ultimately, however, such statements provide only partial intellectual satisfaction. A basic questions remains unanswered: "How does complex, adaptive behavior arise?" It is believed that the model developed in Chapter Seven provides at least a partial answer to this question.

Theoretical Implications of the Model

Adaptation in human populations takes two major forms: change mediated through genetic processes and learning. In processes of the former type, potential changes introduced by mutation are passed on through reproduction, and spread slowly, if at all. In processes of the latter type, potential changes, introduced by invention and diffusion and transmitted through language and demonstration, sometimes spread quite rapidly. Obviously, the model developed in Chapter Seven is applicable only to change through learning. This process assumes that human adaptation is largely a conscious, rational process. For example, Sansai's intricately adjusted system of double- and triple-cropping is the product of sustained effort and experimentation, not unguided, random behavior.

The model has five characteristics which are probably best regarded as "strengths." These features are that the model: accounts for the fact that consensus ordinarily is reached gradually, and sometimes not at all; assumes that ordinary people and statistically oriented scientists approach acquisition of knowledge in essentially the same way; predicts outcomes which are "probabilistic," not "determined"; emphasizes consideration of the "adaptive fitness" of available strategies; and is presented in testable form. Each of these "strengths" will be considered in turn.

Reaching a Consensus

Commonsense formulations of rationality often equate the concept with "choice of that alternative which promises highest yield, payoff, or profit." Acceptance of this definition raises several problems. It implies a high degree of generality, even universality. The rational choice should be obvious to any observer. However, this is not true. Desirable innovations are not adopted immediately and generally, but piecemeal and over a period of time. Why? Traditional explanations emphasize the existence of non-material, social goals and "conservatism." Our model implies a different explanation.

Acceptance of *khaaw kaseet* at Sansai was manifested as a trend, not as a sudden, general shift. Had local farmers carried out a survey, there might have been general agreement concerning the superiority of the new variety. But there is little reason to assume unanimity in this matter. Farmer A collects data from B, C, and D, while D collects information from A, E, and F. It is unlikely that the two will evaluate a new, little known variety in the same way. However, after several seasons, individual evaluations will converge, and a general evaluation will emerge. When one of the rice varieties is new, one should not be surprised if A decides "rationally," to retain the old one. They have sampled different experiments and we can expect initial evaluations to vary somewhat.

Another factor must be considered. It would be less than rational for any decision-maker to concern himself only with general performance. At

Sansai, some farmers did very well with the established rice variety in 1970. It is not obviously "rational" for a farmer who achieves ninety buckets per *rai* with the established variety, to switch to the new one from which his neighbors are receiving only sixty or sixty-five buckets per *rai*.

It is a strength of the model that it predicts that changes will be manifested as trends, not sudden mass movements. It does so because it is based on the assumption that rational decision-makers, even in a single village, work with different information, and within slightly different environmental constraints.

Ordinary People as "Statisticians"

The concept that ordinary people adopt a "statistical" approach to information gathering is not new in anthropology. In 1961 Erasmus expressed his ideas on "frequency interpretation and probable knowledge":

> *Cognition derives its dynamic quality from frequency interpretation--the potentiality to synthesize experiences into predictive generalizations.* At the preindustrial, non-specialized level, frequency interpretation is based entirely upon casual observation, and spectacularity is important for acceptance of innovations. With specialization, technical observations increase and frequency interpretations multiply tremendously, though they tend to be less inspectional. Communication and acceptance of these frequency interpretations becomes increasingly dependent on indirect experience and vicarious observations, and therefore on the literacy and knowledge of the recipients (1961:31, italics added).

Although Erasmus does not develop an explicit model, it is clear that his view of human behavioral evolution is closely allied to ours. It seems likely though, that he attributes undue importance to "spectacularity," and too little importance to the ability of illiterate and semi-literate populations to use data vicariously. In a 1970 publication, Naroll adopts a similar position:

> Most people know in large measure what they like and what they dislike. The must and in fact they do live their lives on the basis of their judgement--usually intuitive--of the probabilities, of the odds: first, the odds that their perceptions are accurate, or inaccurate; second, the odds that their theories about everyday living--most of which come to them via their culture of course--are correct or incorrect.
>
> I argue then that all people are intuitive mathematical statisticians and Popperian logicians. *All people are constantly constructing theories, testing them against reality, and forming conclusions in their minds about the probabilities of their general theories* (1970:27, italics added).

It is a strength of our model that it is based on similar assumptions about human information gathering and behavior. It places these assumptions in an orderly form, and tests them on a collection of specific data. Sansai residents, like statisticians, are assumed to incorporate procedures for sampling, computation, and choice into their daily behavior.

Probabilistic Generalizations

When applied to a set of data, our model predicts a range of possible outcomes. The result of carefully specified conditions is not determined, instead it is expressed as a set of probabilities. The probabilistic nature of the model is "built in." It is introduced with the sampling procedure, under which it is assumed that essentially random events control the information gathering process. Intuitively satisfying, it emphasizes the "chance" nature of daily interactions in village life. The model is intuitively satisfying at another level for it provides some insight into the problems involved in predicting human behavior. The probabilistic nature of this model can be reduced only by developing a model incorporating all factors which control movement around the village area and result in contact with other farmers and exchange of yield information. For the time being, the assumption of randomness appears to be an adequate substitute.

"Adaptive Fitness" of Available Strategies

Our model developed in Chapter Seven provides for the careful examination of the "adaptive fitness" of established and newly available strategies. It is assumed that Sansai farmers carefully consider costs and benefits associated with available rice varieties under *local* conditions. In doing so, they are considering the adaptive fitness of available strategies, both established and "innovative".[1] The concept can be used comparatively. For example, if A costs less and provides greater benefits than B, it possesses greater adaptive fitness. The model can also be used absolutely, e.g. if costs outweigh benefits an innovation lacks adaptive fitness.

In published research on diffusion of innovations adaptive fitness has received little emphasis. In *Communication of Innovations* Rogers and Shoemaker have summarized the results of more than 1000 diffusion studies and attempted to construct appropriate theories. They offer lengthy discussions of "perceived attributes of innovations", "adopter categories", "opinion leadership", "the change agent", and "communication channels" (1971: viii-xii). But their consideration of the significance of relative costs and benefits under local conditions is superficial and disappointing. They emphasize a wide range of *secondary* causal factors and ignore the *primary* factor--adaptive fitness. Some of the factors considered probably are essential to an exhaustive explanation of innovative behavior. However, their causal significance cannot be usefully evaluated except in conjunction with basic data on adaptive fitness.

An example provided by Rogers and Shoemaker illustrates this point. Early in their first chapter they provide an explanation of "Water-Boiling in a Peruvian Village". They offer a brief description of the village and certain public health programs of the Peruvian government. Then they describe three case histories in which Nelida, a public health nurse, works with local women. Each woman is given a different label--"custom oriented", "persuaded", and "rejector". In the paragraph following the third case history we learn that:

> Several housewives, particularly those of the lower social class, are rejectors because they have neither the time nor the means to boil water, even if they are convinced of its value.

> These women lack time to gather firewood and to boil water. The poor cannot afford the cost of fuel for water-boiling and the wives often work as field laborers beside their husbands, leaving them less time to boil water for their families (Rogers and Shoemaker 1971:5).

The above quotation strongly suggests that water boiling lacks adaptive fitness in the village described. The labor and monetary costs apparently are beyond the means of most local households. Yet the authors proceed to an explanation of "why water-boiling failed" in which costs and adaptive fitness are totally ignored. Instead we read that:

> Only the least integrated individuals risk defying community norms on water boiling. . . . Nelida's failure demonstrates the importance of reference group influences in the adoption and rejection of an innovation. . . . The marginal Mrs. B adhered to the standards of a cosmopolite reference group outside the local community. . . . The way that potential adopters view the change agent affects their willingness to adopt his ideas (Rogers and Shoemaker 1971:5-6).

These authors insist on devising complex cultural, social, and psychological explanations while ignoring the apparent lack of adaptive fitness of the innovation. In doing so, they have left out the most important element in any scheme for explaining innovation. Later in the book, when the authors develop an explicit model of the process, they continue to emphasize personality and social characteristics of the potential innovator rather than his ability to pay. They emphasize the communication process, but largely ignore other aspects of the environment within which an innovation is accepted or rejected. They emphasize the *perceived* characteristics of innovations and argue for careful study of attitudes. Yet they ignore the potential value of measuring costs and benefits of the innovation (Rogers and Shoemaker 1971:99-133). In sum, these authors emphasize complex social, cultural, and psychological explanations and place themselves in the dubious position of asserting that "tradition" or "conservatism" prevents adoption of some innovation when simpler explanations, based on the material self-interest of the innovator, are available.

The decision model developed in Chapter Seven emphasizes adaptive fitness, and virtually ignores the other factors considered by Rogers and Shoemaker. This is not the result of an oversight. It is the result of deliberate selection of a situation--choosing rice varieties--in which cultural and social variables can justifiably be considered to "hold constant".

Testing the Model

Our model is testable. A set of data which would invalidate it can easily be imagined. For instance, given the yield figures for early rice at Sansai in 1970, the model would clearly be inadequate if the overwhelming majority of farmers had planned to plant *khaaw lyang nauj* in 1971. The fact that the great majority planned to plant *khaaw kaseet* provides the model with support which is at least moderate. Much of the controversy concerning the economic motives of non-Western farmers has not been carried on at the level of hypothesis- and model-testing. Discussion instead has consisted of exchanging anecdotes. This approach is epitomized in Kusum Nair's books, *Blossoms in the Dust* and *The Lonely Furrow*.

Nair is a journalist by occupation, and her approach to her subject is clearly journalistic. *Blossoms in the Dust* evaluates "the human factor in Indian development" through a series of carefully collected, well-written anecdotes. *The Lonely Furrow* applies the same technique to farmers in the United States and Japan. For both books, Nair traveled thousands of miles, and collected hundreds of interviews. But her evaluation of voluminous research materials is essentially journalistic. She accepts her informants' statements of fact and opinion at face value, providing little or no background data. The following passages exemplify her approach:

> It is a natural assumption, made mainly by those who have never known real poverty, that a poor man can never be in doubt about his problems. He does not have enough for what economists would define as his "absolute" needs, such as food, clothing and housing, and therefore he must want more. Depending upon one's philosophy of life, it is said, one may not be inclined to exert oneself for luxuries. Beyond a certain stage of affluence one may not even be clear as to precisely what one needs. But it should not be necessary to tell a hungry man that he requires food and that he must work to obtain it.
>
> Yet Balappa, a landless agricultural labourer in Gangwati *taluk* of the State of Mysore, has returned from work though the time is only noon. He was fortunate to get work today. But now he is reclining against the wall of his hut smoking a *bidi*.
>
> "How is it you are back so early? Are you not going to work for the rest of the day?"
>
> "No," is the brief reply.
>
> "Why?"
>
> "I am tired."
>
> He is a young man; has a wife and a child and one cow.
>
> "How much have you been paid for a half a day's work?"
>
> "One and a half *seers* of *jowar*."
>
> "But are you able to get work all the year around that you can afford to take it so easy?" I persist.
>
> "No."
>
> "Then what do you do when you have no work?"
>
> "Sit at home and borrow and eat. What else can I do?"
>
> "Why don't you work for the whole day when work is available and save for lean times?" I ask again, unable to restrain myself from offering gratuitous advice.
>
> "Oh, I have to get grass for my cow. If I work in the afternoon as well, when will I get the grass for the cow?", he says, throwing away the still smoking stump of the *bidi* with a gesture of dismissing the debate.
>
> It is not food alone that Balappa lacks for many days in the year. His clothes are torn and dirty, and his house is a small one-room mud hut with a thatch which his family shares with the cow. But he relaxes by the time it is noon and, of his own choice, does not go back to work even when it is available. He possesses nothing but poverty; yet a millionaire could not care less (1962:46-47).

Nair's interpretation of Balappa's behavior is obvious. Unlike some of his neighbors, he is not concerned with the material welfare of his family. That is the "human factor" in agricultural development; some individuals are rational, but many are not. Perhaps so, but Nair's data are clearly inadequate to support her position. No cited case is backed by sufficient information (e.g. on illness, household budgets, soil quality, availability of credit, good and bad fortune, etc.) to justify any judgment concerning economic motives. Perhaps Balappa was ill on the day of the interview. Perhaps it is more profitable to feed the cow than to work in the afternoon. Perhaps he had been fired, and was ashamed to admit it. Perhaps he simply lacked the energy to work through a long, hot afternoon. Perhaps his afternoon wages would have been claimed immediately by his creditors. Anecdotes seldom contain sufficient personal or ecological data to prove or disprove the reporter's general impression of economic irrationality.

It is a strength of our model that it is testable. It takes the controversy outside the *cul de sac* of the exchange of opposing anecdotes.

The Decision Model in Diachronic Perspective

Our model is synchronic and narrowly focused. It applies to decisions made for a single cropping season during one year. This particular season was chosen because it presents few problems in assigning a maximization index. It was also selected because a dramatic shift took place in varietal choice during and following the research period. In this limited case, the model appears to be accurate. However, if it is to be applied in other situations, the model must be generalized. Ideally, it should be applicable to long range, broadly focused decision processes (of the type studied by Behrman). With slight modification, the tentative model in Chapter Seven can be appropriately generalized.

In synchronic form, the model incorporates three procedures:

SAMPLING
↓
COMPUTATION
↓
CHOICE

Time depth can be added simply by assuming that the procedures are cumulative, and iterative, as follows:

SAMPLING
↓
COMPUTATION
↓
CHOICE

In other words, it is assumed that collected data from all previous years are used in comparing crops or varieties. In doing so, it is unnecessary to alter sampling or choice procedures. It is reasonable to assume that each farmer gathers data from the same number of friends, relatives, and neighbors each year, and chooses between varieties in the same manner. However, the amount of data on established varieties increases annually, and occasionally, new varieties come under consideration. For this reason the computation procedure requires slight modification so that it is more general in form.

Modification of the model is accomplished by assuming that as sample data collected over several years proceed toward some arbitrary limit (for example, data from fifty individual harvests over several years) the farmer acquires a "general" knowledge of varietal potential. As individual samples increase in size, individual assessments will converge, and after several seasons, clearly inferior varieties will disappear.

In Chapter Seven the model was applied to a single dimension of choice for a single crop. However, in addition to variety, the farmer makes significant choices concerning planting data, use or non-use of fertilizer or tractor, etc. It is assumed that each of these choices can be modeled separately and simultaneously. For a single crop, all such decisions are directed toward maximization of the same index. For example, for the early rice crop at Sansai, the farmer chooses planting date, variety, etc., *all* with the intention of maximizing yield. It is assumed that all such simultaneous decisions are made rationally.

The maximization index chosen for early rice production at Sansai is yield, expressed in kilograms or buckets per *rai*. For cash crops, assignment of an index may be more complex. A variety which provides inferior yield may receive a superior price. Late harvested rice varieties may bring less than early harvested types. Some varieties are subject to spoilage, while others are not. For commercial crops (e.g. watermelons, garlic, soybeans, and peanuts) at Sansai indices are more appropriately expressed in terms of expected profit.

Implications of the Model for Governmental Policy

Schultz (1963) has emphasized the essential rationality of "traditional" farmers. Short-range, narrowly focused data from Sansai, and Behrman's (1968) long-range, broadly focused data from central and northeastern Thailand lend substantial support to Schultz's position. The policy implications of traditional farmer rationality are substantial.

Schultz argues that the emphasis of many agricultural development programs has been wrong. He states that traditional farmers require governmental services in the areas of research, information, and education. No matter how rational, individual farmers in all parts of the world lack the ability to do controlled research on improved varieties, fertilizers, herbicides, insecticides, planting dates, etc. Responsibility for definition of new strategies necessarily falls to governmental units. Similarly, farmers cannot utilize strategies of which they are unaware. Responsibility for dissemination of research necessarily falls on governmental units. Schultz also argues that education of farmers is an important aspect of agricultural development. Literate decision-makers are in a position quickly to accept innovations. Farmers who have skills in arithmetic and bookkeeping are able quickly and precisely to evaluate innovations.

The model constructed from Sansai data suggests the need for careful attention to attainment of "critical density" in innovative programs. For example, introduction of *khaaw kaseet* in the Sansai area was dependent on a government-supported program of free seed distribution. If it is true that individual assessments of innovations are based on sample data, then the government must carefully assess the "critical density" for materials distributed.

If the single household which experimented with *khaaw kaseet* within Sansai in 1969 had met with significant failure (e.g. the thirty-six buckets per *rai* which another household attained in 1970), it is probably safe to assume that the new variety would have met with substantial resistance at the outset. Its acceptance could have been delayed for several years. This assessment had nothing to do with peasant conservatism, but assigns considerable importance to attainment of critical density. In most cases, five or ten experimenting households are probably sufficient in an area the size of Sansai. More would be unnecessarily expensive, and fewer would jeopardize the innovative program in some areas. A single bad experience will be sufficient to convince farmers of the inferiority of an innovation *if* farmers have only one case upon which to make evaluations.

If it is true that traditional farmers are essentially rational, then it is crucial that bureaucratic organizations responsible for agricultural innovation retain sufficient flexibility to reevaluate programs which prove to be unsuccessful. Careful evaluation will show that many are economically inappropriate in specific areas, e.g. that established methods return higher profits to land or labor. An innovation program directed toward rational farmers will have the following elements: strong emphasis on research (or "development of new strategies"); imaginative programs for making available information on innovations which represent genuine economic improvement; dissemination of skills in reading and bookkeeping which will aid farmers in learning about innovations and evaluating them; careful attention to attainment of critical density in making field trials; and built-in procedures for reevaluation of failing programs.

Our model presents Sansai farmers as self-interested, flexible, economic maximizers, eagerly adopting innovations which promise economic advantage, and readily discarding crops, varieties, tools, and techniques which are no longer superior. This model, and analogous models applied to similar situations, probably are sufficient to account for present adaptive patterns at Sansai.

GLOSSARY

Adaptive fitness. A useful index in evaluating possible or available innovations. The concept can be used both comparatively--if A costs less and provides greater benefits than B it possesses greater adaptive fitness--and absolutely--if costs outweigh benefits an innovation lacks adaptive fitness.

Adaptive strategy. A system of "cognition and behavior maximizing the well-being and minimizing the hazards to survival of human populations" (Shimkin, 1966:38).

Amphoe. See District.

Baht. Primary unit of Thai currency equal to U.S. 5 cents.

Basket (*thaang*). An approximate measure of volume used in agriculture. Sansai baskets usually have a capacity of about 30 liters.

Bau ngaam. Northern Thai for "ugly" or "not pretty." Commonly used to describe a stunted, unhealthy stand of plants.

Bran A by-product of rice milling consisting of the plant layers closest to the white endosperm. The "husk" or outer layer provides a less valuable by-product.

Bucket (*tang*). A standardized measure of volume used in Thai agriculture that is equivalent to 20 liters.

Bunds. The narrow mud walls which separate rice plots and maintain standing water on the plots.

Changwad. See Province.

CUSSRI. Chulalongkorn University Social Science Research Institute.

Decision against Nature. A special case of decision-making under uncertainty in which the decision-maker "copes" with uncertainty by treating partial knowledge as if it were complete.

Diachronic. Refers to an analysis or description concerned with events taking place over an extended period of time.

District (*amphoe*). An administrative unit of the Thai government one level below the province.

Early rice (*naa dau*). This is the Sansai rice crop that ordinarily is planted in nurseries in February-March and harvested in June-July.

Game Theory. A set of mathematical techniques for maximizing some goal-index in situations in which uncertainty can be reduced to certainty by assuming that the goals of opposing "players" can be ascertained in advance.

Garden crop (*suan*). At Sansai these non-rice crops are planted in November-January and harvested in February-April.

Glutinous rice. Rice varieties commonly raised in northern and northeastern Thailand and Laos that cohere in a sticky mass when cooked and have chemical properties different from ordinary or non-glutinous varieties.

Household. A group of people (often, but not necessarily, belonging to a single nuclear or extended family) who sleep under one roof and eat "from the same pot."

Khaaw kaew dau. A rice variety planted at Sansai for the late rice crop. Sometimes abbreviated in this monograph as KKD.
Khaaw kam phaaj. A rice variety planted at Sansai for the late rice crop. Sometimes abbreviated in this monograph as KKP
Khaaw kaseet. A rice variety planted at Sansai for the early rice crop. Sometimes abbreviated in this monograph as KK.
Khaaw lyang nauj. A rice variety planted at Sansai for the early rice crop. Sometimes abbreviated in this monograph as KLN.
Khaaw lyang San Pratong. A rice variety planted at Sansai for the late rice crop. Sometimes abbreviated in this monograph as KLS.
Khon myang. Native residents of northern Thailand.

Lamyai. See Longan.
Late rice (*naa pii*) At Sansai this is the rice crop that ordinarily is planted in nurseries in June-July and harvested in November-December.
Linear programming. A set of mathematical techniques for maximizing some goal-index in the face of known costs and no risks.
Longan (*lamyai*). A fruit, raised in northern Thailand, that is similar to the lychee.

Maximization. Behavior oriented toward improving or maintaining the level of some goal-index.

Naa dau. See Early rice.
Naaj. Northern Thai title. As presently used in rural areas it is roughly equivalent to Mr.
Naa pii. See Late rice.
Ngaam. Northern Thai for "pretty." Commonly used in agriculture to describe a healthy stand of plants.
Nutritional density. An index of human population density in which population size is considered in relationship to area of cropped land.

Paddy, padi. Unhusked or unmilled harvested rice.
Photoperiod sensitive. A descriptive term applied to plants (including many rice varieties) which ripen or flower in response to specific patterns of day length, and thus can be expected to mature only during one, restricted period of the year.
Productivity. An agricultural index in which the weight, volume, or cash value of some commodity is measured against the amount of labor required to produce it.
Province (*changwad*). Major administrative units of Thailand.

Rai. A unit of land equal to 0.16 hectare or 0.39 acre.

Suan. See Garden Crop.
Subdistrict. An administrative unit of the Thai government one level below the district. This unit sometimes is translated into English as "commune."

Synchronic. Refers to an analysis or description concerned with events taking place at a single point in time.

Tambon. See Subdistrict.
Tang. See Bucket.
Technologically simple. A phrase used to describe rice farmers who use simple tools and rely primarily on human, animal, and direct solar energy sources.
Thaang. See Basket.

Wat. A Buddhist religious center which combines the functions of a temple and monastery.

Yield. An agricultural index in which the weight, volume, or cash value of some commodity is measured against the area of land required to produce it.

NOTES

Chapter One

[1] An acre of agricultural land at Sansai cost $600-$1000 in 1969-70, a price relatively close to the $1000-$1500 paid for farm land in McLean County, Illinois, during the same years. However, hired men in McLean County could expect to earn at least $15-$20 per day for their labor, while agricultural laborers at Sansai were receiving 25¢-50¢ per day. Sansai land prices are *extremely* high in relationship to labor costs.

Chapter Three

[1] The major anomalies are in age groups 20-29 and 50-59 years. The small size of the former group probably can be explained by the smallpox epidemic which followed World War II, while the small size of the latter group probably can be explained by the flu epidemic of 1918-19.

[2] Home-grown rice, consumed in most households for at least part of each year, was assigned the prevailing retail value of one *baht* per liter.

[3] Households A, B, H, and I all had members who spent part of the year under discussion away from the village, either working or attending school. The son of Household Head D acquired a new wife during the last month of the survey period, while Household Head G was ill and unable to work for several months.

[4] In Sansai, wood and charcoal fires are used exclusively for cooking. Even in the cool season they are never maintained for the warmth provided.

Chapter Four

[1] For additional information see Sharon Kay Calavan (1974).

[2] A similar interpretation of agricultural choice is applied by Nash to Burmese villagers (1965).

[3] These categories are used by Shimkin (1956).

Chapter Six

[1] The amounts shown are incremental over a predicted yield without fertilizer of 375 *baht* per *rai*.

[2] "Total expenditure" includes fertilizer cost, as well as costs of tractor rental and labor.

[3] Fertilizer expenditures were measured in *baht* spent, and yield in number of fruits produced. Neither of these is a particularly satisfactory measure, but they are the only relevant data easily collected in extensive, survey-type research.

[4] In addition to fertilizer, this item also includes the cost of seed, insecticide, hired labor, transportation, and "miscellaneous."

Chapter Seven

[1] Linear programming provides one example of a "decision under certainty."

[2] Roulette provides an example of a "decision under risk."

[3] Crop choice provides an example of a "decision under uncertainty."

[4] This estimate is based on the following known and estimated figures. Moerman reports that Ban Ping consists of 120 households, and it is here estimated that the average household has 5.0 members, and controls 0.5 *rai* of houseland (1968).

[5] Davenport's Jamaican Fishing: A Game Theory Analysis, is one of the earliest attempts in anthropology to utilize the Game Theoretic model (1960). In recent years two major critiques of the article have appeared (Read and Read 1969; Kozelka 1969). Both argue that the fact that the model seems to "work," that actual strategies and those prescribed by game theory very nearly coincide, is epiphenomenal. The authors of both critiques are convinced that Davenport is wrong to model Nature as a self-interested, opposing player as Game Theory requires.

[6] This estimate is based on intuition and crude observation. Later it will be demonstrated this arbitrarily chosen sample size, or any other, has little effect on values predicted by the model.

[7] Computer simulation was rejected because the writer calculated that program writing would occupy more time than the less sophisticated method used. Yield data, varietal name, and household identification were recorded on note cards, one card for each household. The "sampling household" was designated by drawing a card from the deck, and setting it aside. The deck was then shuffled and cut several times, and twenty cards (constituting a "random sample") were drawn from it. The data were recorded. The process was continued until each of sixty-one households had drawn a total of twenty "random samples."

[8] Increased sample size decreases the probability that the farmer will collect *no* data on the new variety, and thus plant *khaaw lyang nauj* out of ignorance.

[9] The discussion is restricted to the effect of procedural alteration on computed value for the established variety. During the 1971 period, alteration of computation method for the new variety will have little effect on predicted outcomes.

10 In all cases, the range is calculated to extend three standard deviations above and below the mean.

Chapter Eight

1 To be sure, consideration of costs and benefits is relatively simple. A simple case was chosen deliberately. The costs--in cash, labor, equipment--are assumed to be virtually identical for new and established rice varieties. Thus, considerations of adaptive fitness tend to focus on mean yields and variations around those means.

Appendix 1

CUSSRI Socio-economic Profile Schedule

Household No.	Name (a)	Age (b)	Sex (c)	Place of Birth (d)	Relationship to head of Household (e)	Martial Status (f)	Level of Educ. (g)	Work Status (h)
1, etc.								

List household members who are in the labor force

Household Member No.	Name	Type of Occupation (Industry) (i)	Main Occupation (j)	Subsidiary Occupation (k)	Status in Main Occupation (l)	Status in Subisidiary Occupation (m)	Months Worked on Family Farm (n)	Farm Work Capacity (percent) (o)
1, etc.								

Children of household head permanently away from home

Household Member No.	Name (a)	Sex (b)	Age (c)	Level of Education (d)	Age (e)	Level of Education (f)	Type of Occupation (g)	Main Occupation (h)	Where now Living (i)	Reason for Leaving Home (j)
1, etc.										

Appendix 1 (Continued)

Farm size and land use

	Owned (a)	Ownership Title (b)	Rented In (c)	Rented Out (d)	Used Without Ownership Title (e)	Operated (f)	Percent of Agricultural Land (g)
1 Arable land							
2 Land under tree crops							
3 Pasture land							
4 Agricultural land (substract total 1+2+3)							
5 Wood land							
6 Other land							
7 Total land (4+5+6)							
8 Land under wet season irrigation							
9 Land under all year irrigation							
10 Percent of operational holding							

(h) Source of irrigation water for all year irrigation

Appendix 1 (Continued)

For each member of the household who engaged in own-account non-farm work that has been a source of income during the 12 months of the last agricultural year.

Household Member No. (a)	Name (b)	Job No. (c)	Type of Work (d)	Place of Job (e)	Distance from Home (f)	How often Returned Home (g)	Months Spent on Job (h)	Percent of Work time Spent on Job (i)	Total Income Received From this Job (j)	How much Given to Household (k)
1, etc.										

For each member of the household who took full or part-time employment during the 12 months of the last agricultural year.

Household Member No. (a)	Name (b)	Job No. (c)	Type of Work (d)	Place of Job (e)	Distance from Home (f)	How often Returned Home (g)	Months Spent on Job (h)	Percent of Work time Spent on Job (i)	Total Income Received From this Job (j)	How much Given to Household (k)
1, etc.										

Appendix 1 (Continued)

Present Crops Planted

	Wet Season Crop (a)	No. of *rai* (b)	Percent of Arable Land (c)	Most Recent Year (d)		Average Last 5 yrs. (e)		Total Production Most Recent Year (f)	Units Retained on Farm (g)	Units Sold of Bartered (h)	Where sold (i)	To Whom Sold (j)
				Local Unit	kg.	Local Unit	kg.					
1												
2												
3												
4												
5	Fallow											
	Total											
	Dry Season Crop											
1												
2												
3												
4												
5	Fallow											
	Total											

Appendix 1 (Continued)

Modern equipment used

Complete the following check list of equipment

Modern equipment	Owned	Hired	Borrowed
Tractor			
Hand tractor			
Insect pest sprayer			
Power pump			
Harvest			
Thresher			
Home rice mill			
Truck			
Motorcycle			
Boat motor			
Electrical appliances			
Other, specify			

Appendix 1 (Continued)

Livestock inventory

Kind of Livestock (a)	Work Animals (b)	Farm Animals (c)	No. sold (d)	Average Price Per head (e)	Where Sold (f)	To Whom Sold (g)	No. Stolen or lost (h)	No. Died (i)
1 buffalo								
2 cattle								
3 hogs								
4 chickens								
5 ducks								
6 geese								
7 other, specify								

Appendix 2 *Household Daily Schedule**

Activity

1. Name________________

etc.______________________ ___ ___ ___

2. Name________________

etc.______________________ ___ ___ ___

3. Name________________

etc.______________________ ___ ___ ___

4. Name________________

etc.______________________ ___ ___ ___

5. Name________________

etc.______________________ ___ ___ ___

*Respondents were asked to record numerous additional data, but none was relevant to this research project.

Appendix 3 *Household Expenditures, 1969*

House#______
Village#____

I. *Agriculture and Food Production*

Did you buy any of the following?

Seeds or seedlings _____
Fertilizer _____
Insecticide _____
Insecticide sprayer _____
Water pump _____
Agricultural land _____
Oxen _____
Buffalo _____
Pigs _____
Poultry _____
Rice straw (for garlic) _____
Animal feed (all types) _____
Plastic bags (for melons) _____
Hoes _____
Knives _____
Ox-cart _____
Ax or hatchet _____
Harvest basket _____
Plow _____
Harrow _____
Watering cans _____
Fishing equipment _____
Other items useful in farming _____

SUB-TOTAL A _____

Did you pay rent for any of these?

Tractor _____
Hand tractor _____
Water pump _____
Draft animals _____
Land (garden crop) _____
Spraying equipment _____
Other items _____

SUB-TOTAL B _____

Repair of tools or equipment _____
Medicines for livestock _____
Kaakaan (owned or rented land) _____
Well digging _____
Other services _____

SUB-TOTAL C _____

TOTAL _____

II. *Household Necessities*

Did you buy any of the following?

Flashlight _____
Pressure lamp _____
Other lamp _____
Candles _____
Lamp fuel _____
Alcohol _____
Batteries _____
Mats (all types) _____
Baskets (all types) _____
Clock _____
Water pots _____
Cooking pots _____
Dishes and glasses _____
Knives, spoons, forks _____
Chairs _____
Tables _____
Mattresses _____
Sheets and blankets _____
Cabinets _____
Washing powder _____
Photos, illustrations _____
Stamps, stationery _____
Pens, pencils _____
Buckets _____
Dippers _____
Other items _____

SUB-TOTAL A _____

Appendix 3 (Continued)

Did you spend money on any of these?

Repair of clock, lamp, pots, etc. ______
Servant's salary ______

SUB-TOTAL B ______

TOTAL ______

III. *Housing and Other Structures*

Did you build or have built any of these?

House ______
Bathhouse or privy ______
Grain storage bin ______
Pig pen ______
Fence ______
Shop or business ______
Other structure ______

SUB-TOTAL A ______

Did you spend money on any of these?

Material to repair any structure ______
Labor to repair any structure ______

SUB-TOTAL B ______

Did you purchase any of these?

House land ______
Other land (not for farming) ______

SUB-TOTAL C ______

TOTAL ______

IV. *Transportation*

Did you purchase any of these?

Bicycle ______
Motorcycle ______
Truck ______
Automobile ______
Gasoline and oil ______
Tires ______

SUB-TOTAL A ______

Did you pay for any of these?

Repairs for bicycle ______
Repairs for motorcycle ______
Repairs for truck ______
Repairs for automobile ______

SUB-TOTAL B ______

Did you spend money on the following?

Riding buses, taxis, etc. ______
Transporting goods in buses, taxis, etc. ______

SUB-TOTAL C ______

TOTAL ______

V. *Business Expenses*

Did you buy any of the following?

Rice mill ______
Tractor ______
Hand tractor ______
Gasoline and oil for rice mill, tractor, etc. ______

Appendix 3 (Continued)

Tools of trade _____
Other equipment _____
Supplies _____

SUB-TOTAL A _____

Did you spend money on any of these?

Repairs to tools and equipment _____
Worker's salaries _____

SUB-TOTAL B _____

TOTAL _____

VI. *Finances*

Did you pay out money for any of these?

Land taxes _____
Business taxes _____
Loan payments _____
Life insurance payments _____
Lending out money _____
Assessments for local improvements _____
Legal fines _____

TOTAL _____

VII. *Medical Expenses*

Did you spend money for any of these?

Medicines _____
Injections _____
Consultations, house calls _____
Hospital care _____

TOTAL _____

VII. *Education*

Did you spend money for the following?

Tuition and fees _____
Books _____
School supplies _____
Room and board _____
Transportation (home to school) _____

TOTAL _____

IX. *Religion and Merit-Making*

Did you make merit in any of these ways?

Feeding monks on ordinary days _____
Tham bun on *waansiin* _____
Gifts for weddings _____
Gifts for funerals _____
Gifts for religious construction _____

TOTAL _____

X. *Entertainment*

Did you spend money on any of these?

Radio _____
Tape recorder _____
Tapes _____
Record player _____
Records _____
Repair of radio, record player, recorder _____
Newspapers, magazines _____
Gambling losses _____
Children's toys _____
Cigarettes and makings _____
Gun _____

Appendix 3 (Continued)

Ammunition ______
Repair of gun ______
Liquor ______
Tickets for movies, shows, temple festivals ______

TOTAL ______

XI. *Clothing and Personal Grooming*

Did you spend money on the following?

Watch ______
Jewelry ______
Cosmetics ______
Toothpaste, toothbrush, soap, combs, etc. ______
Clothing, shoes (adult males) ______
Clothing, shoes (adult (females) ______
Clothing, shoes (boys) ______
Clothing, shoes (girls) ______
Haircuts for males ______
Hairdressing for females ______

TOTAL ______

XII. *Food*

Did you buy the following items?

Milled rice ______
Unmilled rice ______
Vegetables ______
Fruits ______
Meat ______
Fish ______
Eggs ______
Naampaa ______
Salt ______
Naamphik ______
Sugar ______
Cooking oil ______
Paahaa, kapii ______
Khanom ______
Kaeng (prepared) ______
Restaurant meals ______
Rice milling ______
Other food items ______
(Amount of home-grown rice used @1 *baht* per liter) ______

TOTAL ______

XIII. *Double Check*

How much rice do you use?

Per day ______
Per week ______

How much *kab khaaw* ("with rice" food) do you use?

Per day ______
Per week ______

TOTAL ______

GRAND TOTAL ______

Percent spent on food ______

Appendix 4 *"Early Rice" Crop Survey*

I. Operator of the Land

1. Name ________________________ Village #______ House #______

2. Yields

Rice variety	Area	Production
____________	____	__________
____________	____	__________
____________	____	__________
____________	____	__________

3. Labor Force (Household Members Who Worked on This Crop)

_____________	_____________
_____________	_____________
_____________	_____________
_____________	_____________

II. Pattern of Landholding

1. The operator owns the land_________ The land is rented_________

2. Rental arrangement Cash rent_____ Half shares_____ Other_____

III. Fulfillment of social obligations through corvee labor.

1. Roadwork_______________days
2. Work at temple_________days
3. Work at cremation ground___days
4. Work at school_____________days
5. Work on irrigation system______ days

IV. Timing of rice crop

1. Date that nursery was planted______________________________
2. Date that seedlings were transplanted______________________________
3. Date that harvesting was initiated______________________________

* Respondents were asked to provide additional information, but items not directly relevant to this research have been deleted.

Appendix 4 (Continued)

V. Inputs of Labor and Equipment

Task	Labor and Equipment Used														
	Household Labor	Hired Labor		Buffalo		Tractor		Hired Tractor		Truck		Ox-cart		Irrigation Pump	
	days	days	฿	days	฿	days	฿	days	฿	days	฿	days	฿	days	฿
1. Preparation of nursery															
2. Sowing nursery															
3. Watering nursery															
4. Plowing															
5. Harrowing															
6. "Cleaning" the field															
7. Pulling, cutting, and carrying seedlings															
8. Transplanting															
9. Weeding															
10. Irrigation															
11. Reaping															
12. Threshing															
13. Carrying paddy to storage															
14. Sell rice															

Appendix 5 *Garden Crop Survey*

A. Name_______________________________ House #__________ Village #__________

B. Amount of land owned_____ *rai* Owned land worked_____ *rai* Rented out _____ *rai* Unused land_____ *rai* Rented land_____ *rai* Rental Arrangement_________________

C. Number of planted in garden crops

Garlic_____ *rai* Watermelons_____ *rai* Soybeans_____ *rai* Peanuts_____ *rai*

D. How many years have you raised garden crops?_________________________

E. Tell me about the crops. Have you changed from previous years? Why? Why not?____

____

F. Do you need to guard your crops in the field?_________________________

____

G. Do you ever have problems with thieves, insects, drought, flooding, etc.?____

____

H. Watermelons

I. Which variety?__________________________ How much seed?________________

Price of seed____________________ Where did you get it?________________

J. Total production________________ Price received______________________

K. Cost of fertilizer______________ Tractor rental______________________

Insecticide_____________________ Hand tractor rental________________

Hired labor____________________ Insecticide sprayer rental____________

Appendix 5 (Continued)

Buffalo rental__________________ Truck rental_______________________

Ox-cart rental__________________ (Other expense)____________________

L. Garlic

M. Source of seed_____________________ Price of seed___________________

N. Total production___________________ Price received__________________

O. Have you sold your garlic yet? If so, explain the sale arrangement. If not, why not? When the time comes, how will you sell it? Why? Will you keep seed for next year?

____

____

P. Cost of fertilizer__________________ Tractor rental__________________

Insecticide______________________ Hand tractor rental________________

Hired labor___________________ Insecticide sprayer rental____________

Buffalo rental__________________ Truck rental_______________________

Ox-cart rental __________________ (Other expense)____________________

Q. Soybeans

R. Source of seed________________________ Price of seed__________________

S. Total production______________________ Price received_________________

T. Have you sold your soybeans yet? If so, explain the sale arrangement.____

____

U. Cost of fertilizer__________________ Tractor rental__________________

Insecticide________________________ Hand tractor rental______________

Hired labor____________________ Insecticide sprayer rental____________

Buffalo rental______________________ Truck rental______________________

Ox-cart rental______________________ (Other expense)__________________

Appendix 5 (Continued)

V. Peanuts

W. Source of seed____________________ Price of seed________________

X. Total production__________________ Price received_______________

Y. Have you sold your peanuts yet? If so, explain the sale arrangement.____

____

Z. Cost of fertilizer____________________ Tractor rental____________

Insecticide______________________ Hand tractor rental____________

Hired labor__________________ Insecticide sprayer rental__________

Buffalo rental______________________ Truck rental_______________

Ox-cart rental______________________ (Other expense)_____________

Labor Expenditures (Man-days)

Watermelons	Household Labor	Hired Labor	Buffalo	Tractor	Hand Tractor	Sprayer	Ox-cart	Truck
1. Buy seed and fertilizer	_	_	_	_	_	_	_	_
2. Plowing	_	_	_	_	_	_	_	_
3. Digging shallow wells	_	_	_	_	_	_	_	_
4. Making seed containers	_	_	_	_	_	_	_	_
5. Planting seeds	_	_	_	_	_	_	_	_

Garlic	Household Labor	Hired Labor	Tractor	Hand tractor	Buffalo	Sprayer	Ox-cart	Truck
1. Buy fertilizer	_	_	_	_	_	_	_	_
2. Prepare garlic seed	_	_	_	_	_	_	_	_
3a. Plowing	_	_	_	_	_	_	_	_
3b. Dig small ditches	_	_	_	_	_	_	_	_
4. Prepare planting beds	_	_	_	_	_	_	_	_
5. Plant	_	_	_	_	_	_	_	_

Appendix 5 (Continued)

Watermelons Con't.	Household Labor	Hired Labor	Buffalo	Tractor	Hand Tractor	Sprayer	Ox-cart	Truck
6. Digging holes	_	_	_	_	_	_	_	_
7. Watering seedlings	_	_	_	_	_	_	_	_
8. Placing fertilizer in holes	_	_	_	_	_	_	_	_
9. Plant and fill holes	_	_	_	_	_	_	_	_
10. Spread more fertilizer	_	_	_	_	_	_	_	_
11. Watering	_	_	_	_	_	_	_	_
12. Weeding	_	_	_	_	_	_	_	_
13. Collect and grade	_	_	_	_	_	_	_	_
14. Locate a rental truck	_	_	_	_	_	_	_	_
15. Transport and sale	_	_	_	_	_	_	_	_

Garlic Con't.	Household Labor	Hired Labor	Buffalo	Tractor	Hand Tractor	Sprayer	Ox-cart	Truck
6. Prepare rice straw	_	_	_	_	_	_	_	_
7. Spread fertilizer and straw	_	_	_	_	_	_	_	_
8. Irrigation	_	_	_	_	_	_	_	_
9. Weeding	_	_	_	_	_	_	_	_
10. Harvest	_	_	_	_	_	_	_	_
11. Carry garlic from fields	_	_	_	_	_	_	_	_
12. Cutting	_	_	_	_	_	_	_	_
13. Drying	_	_	_	_	_	_	_	_
14. Bundling	_	_	_	_	_	_	_	_
15. Storage	_	_	_	_	_	_	_	_
16. Sale	_	_	_	_	_	_	_	_

Soybeans								
1. Buy seed	_	_	_	_	_	_	_	_
2. Plowing	_	_	_	_	_	_	_	_
3. Digging small ditches	_	_	_	_	_	_	_	_
4. Planting	_	_	_	_	_	_	_	_

Peanuts								
1. Prepare seed	_	_	_	_	_	_	_	_
2. Plowing	_	_	_	_	_	_	_	_
3. Digging small ditches	_	_	_	_	_	_	_	_
4. Planting	_	_	_	_	_	_	_	_

Appendix 5 (Continued)

Soybeans Con't.	Household Labor	Hired Labor	Buffalo	Tractor	Hand Tractor	Sprayer	Ox-cart	Truck
5. Irrigation								
6. Weeding								
7. Harvest								
8. Carry soybeans from field								
9. Preparation for sale								
10. Sale								

Peanuts Con't.	Household Labor	Hired Labor	Tractor	Hand Tractor	Buffalo	Sprayer	Ox-cart	Truck
5. Irrigation								
6. Weeding								
7. Harvest								
8. Carry peanuts from field								
9. Preparation for sale								
10. Sale								

BIBLIOGRAPHY

Bateman, M. J.
1965 Aggregate and Regional Supply Functions for Ghanian Cocoa, 1946-1962. *Journal of Farm Economics* 47:384-401.

Behrman, Jere R.
1968 *Supply Response for Underdeveloped Agriculture: A Case Study of Four Major Annual Crops in Thailand, 1937-1963.* Amsterdam: North-Holland.

Boeke, J. H.
1953 *Economics and Economic Policy of Dual Societies.* Haarlem: H. D. Tjeenk Willink.

Boserup, Ester
1965 *The Conditions of Agricultural Growth: The Economics of Agrarian Change under Population Pressure.* Chicago: Aldine.

Buck, John L.
1930 *Chinese Farm Economy: A Study of 2866 Farms in Seventeen Localities and Seven Provinces in China.* Shanghai: Institute of Pacific Relations.

Burling, Robbins
1968 Maximization Theories and the Study of Economic Anthropology. In *Economic Anthropology: Readings in Theory and Analysis*, edited by Edward E. LeClair, Jr. and Harold K. Schneider:168-187. New York: Holt, Rinehart and Winston.

Calavan, Michael M.
1973 *Cultural, Environmental, and Socioeconomic Factors in Food Use: The Sansai Case.* SEADAG Paper. New York: The Asia Society--SEADAG.

Calavan, Michael M.
1974 *Decisions Against Nature: Crop Choice in a Northern Thai Village.* Ph.D. Dissertation, University of Illinois, Department of Anthropology.

Calavan, Sharon K.
1974 *Aristocrats and Commoners in Rural Northern Thailand.* Ph.D. Dissertation, University of Illinois, Department of Anthropology.

Cancian, Frank
1972 *Change and Uncertainty in a Peasant Economy: The Maya Corn Farmers of Zinacantan.* Stanford: Stanford University Press.

Chandraratna, M. F.
1964 *Genetics and Breeding of Rice.* London: Longmans.

Chang, Te Tzu and Eliseo Bardenas
1965 *Present Knowledge of Rice Genetics and Cytogentics.* Technical Bulletin 4. Los Banos, Philippines: International Rice Research Institute.

Cook, Scott
1968 The Obsolete 'Anti-Market' Mentality: A Critique of the Substantive Approach to Economic Anthropology. In *Economic Anthropology: Readings in Theory and Analysis*, edited by Edward E. LeClair, Jr. and Harold K. Schneider: 208-228. New York: Holt, Rinehart and Winston.

Cunningham, Clark E. and Theodore C. Doege
1969 Illnesses of Saraphi People Treated at Chiang Mai Health Facilities, 1967. MSS.

Dalton, George
1967 Economic Theory and Primitive Society. In *Economic Anthropology: Readings in Theory and Analysis*, edited by Edward E. LeClair, Jr. and Harold K. Schneider: 143-167. New York: Holt, Rinehart and Winston.

Davenport, William
1960 Jamaican Fishing: A Game Theory Analysis. *Yale University Publications in Anthropology* 59:3-11.

Desai, D. K.
1963 *Increasing Income and Production in Indian Farming.* Bombay: Indian Society of Agricultural Economics.

Dumond, Donald E.
1965 Population Growth and Cultural Change. *Southwestern Journal of Anthropology* 21:1231-1257.

Dumont, Rene
1957 *Types of Rural Economy: Studies in World Agriculture.* New York: Praeger.

Erasmus, Charles J.
1961 *Man Takes Control: Cultural Development and American Aid.* Indianapolis: Bobbs-Merrill.

Foster, George M.
1965 Peasant Society and the Image of Limited Good. *American Anthropologist* 67:293-315.

Gallin, Bernard
1966 *Hsin Hsing, Taiwan: A Chinese Village in Change.* Berkeley: University of California Press.

Geertz, Clifford S.
1963 *Agricultural Involution.* Berkeley: University of California Press.

Gould, Julius and William L. Kolb
1964 *A Dictionary of the Social Sciences.* New York: Free Press of Glencoe.

Gourou, Pierre
1960 *The Tropical World: Its Social and Economic Conditions and Its Future Status.* New York: Wiley.

Grist, D. H.
1959 *Rice.* London: Longmans.

Hanks, Lucien M.
1972 *Rice and Man: Agricultural Ecology in Southeast Asia.* Chicago: Aldine-Atherton.

Harris, Marvin
1966 The Cultural Ecology of India's Sacred Cattle. *Current Anthropology* 7:51-66.

Herklots, G. A. C.
1950 Chinese Rice Cultivation. *World Crops* 2:321-324.

Hopper, W. David
1957 *The Economic Organization of A Village in North Central India.* Ph.D. Dissertation, Cornell University, Department of Agricultural Economics.

Joy Leonard
1967 One Economist's View of the Relationship between Economics and Anthropology. In *Themes in Economic Anthropology:* 29-46. London: Tavistock.

Kendall, Maurice G. and William R. Buckland
1971 *A Dictionary or Statistical Terms.* Edinburgh: Oliver and Boyd.

Keyes, Charles F.
1969 Review of *Agricultural Change and Peasant Choice in a Thai Village*, Michael Moerman. *Journal of Asian Studies* 28: 907-908.

King, F. H.
1927 *Farmers of Forty Centuries, or Permanent Agriculture in China, Korea, and Japan.* London: Jonathan Cape.

Kingshill, Konrad
1965 *Ku Daeng: The Red Tomb.* Bangkok; Bangkok Christian College.

Kozelka, Robert
1969 A Bayesian Approach to Jamaican Fishing. In *Game Theory in the Behavioral Sciences*, edited by I. R. Buchler and H. Nutini, 117-125. Pittsburgh: University of Pittsburgh Press.

Lewis, Oscar
1963 *Life in a Mexican Village: Tepoztlan Restudied.* Urbana: University of ILlinois Press.

Luce, R. Duncan and Howard Raiffa
1957 *Games and Decisions.* New York: Wiley.

Meggers, Betty J.
1971 *Amazonia: Man and Culture in a Counterfeit Paradise.* Chicago: Aldine-Atherton.

Mendiola, W. B.
1949 Rice Cultivation in Taiwan. *Philippine Journal of Agriculture* 14:41-83.

Moerman, Michael
1968 *Agricultural Change and Peasant Choice in a Thai Village.* Berkeley: University of California Press.

Moorman, F. R. and S. Rojanasoonthon
1967 *Kingdom of Thailand General Soil Map.* Bangkok: Department of Mineral Resources.

Nair, Kusum
1962 *Blossoms in the Dust: The Human Factor in Indian Development.* New York: Praeger.

Nair, Kusum
1969 *The Lonely Furrow: Farming in the United States, Japan, and India.* Ann Arbor: University of Michigan Press.

Nash, Manning
1965 *The Golden Road to Modernity: Village Life in Contemporary Burma.* New York: Wiley.

Nash, Manning
1968 The Social Context of Economic Choice in a Small Society. In *Economic Anthropology: Readings in Theory and Analysis*, edited by Edward E. LeClair, Jr. and Harold K. Schneider: 311-322. New York: Holt, Rinehart and Winston.

Naroll, Raoul
1970 The Logic of Generalization: Epistemology. In *A Handbook of Method in Cultural Anthropology*, edited by Raoul Naroll and Ronald Cohen: 25-30. Garden City, New York: The Natural History Press.

Oxford Economic Atlas of the World
1965 New York: Oxford University Press.

Pendleton, Robert L.
1962 *Thailand: Aspects of Land and Life.* New York: Duell, Sloan, and Pearce.

Polanyi, Karl
1968 The Economy as Instituted Precess. In *Economic Anthropology: Readings in Theory and Analysis*, edited by Edward E. LeClair and Harold K. Schneider: 122-143. New York: Holt, Rinehart and Winston.

Potter, Jack M.
1976 *Thai Peasant Social Structure.* Chicago: University of Chicago Press.

Random House Dictionary of the English Language
1969 New York: Random House.

Rappaport, Roy A.
1968 *Pigs for the Ancestors: Ritual in the Ecology of a New Guinea People.* New Haven: Yale University Press.

Read, Dwight W. and Catherine E. Read
1970 A Critique of Davenport's Game Theory Analysis. *American Anthropologist* 72:351-355.

Robbins, Lionel
1968 The Subject Matter of Economics. In *Economic Anthropology: Readings in Theory and Analysis*, edited by Edward E. LeClair, Jr. and Harold K. Schneider: 88-100. New York: Holt, Rinehart and Winston.

Rogers, Everett M. and F. Floyd Shoemaker
1971 *Communication of Innovations.* New York: Free Press.

Samuelson, Paul A.
1967 *Economics: An Introductory Analysis.* New York: Free Press.

Schultz, Theodore W.
1964 *Transforming Traditional Agriculture.* New Haven: Yale University Press.

Shimkin, Demitri B.
1966 Adaptive Strategies: A Basic Problem in Human Ecology. In *Three Papers in Human Ecology*, Mills College Assembly Series 1965-1966:45-52. Oakland, California: Mills College.

Silcock, Thomas Henry
1970 *The Economic Development of Thai Agriculture.* Ithaca: Cornell University Press.

Sitiporn Kridikara, H.S.H. Prince
1969? *Aspects of Rice Farming in Siam.* Bangkok: Suksit Siam.

Tanaka, A., K. Kawana, and J. Yamaguchi
1966 *Photosynthesis, Respiration, and Plant Type of the Tropical Rice Plant.* Technical Bulletin 7. Los Banos, Philippines: International Rice Research Institute.

Tax, Sol
1953 *Penny Capitalism: A Guatemalan Indian Economy.* Washington: Institute of Social Anthropology, Smithsonian Institution.

Thailand
1961 *Agriculture in Thailand.* Bangkok: Ministry of Agriculture.

United States Department of Agriculture
1963 *Composition of Foods.* Agriculture Handbook No. 8. Washington: Agricultural Research Service.

Valyasevi, A., S. B. Halstead, and S. Dhanamitta
Studies of Bladder Stone Disease in Thailand. VI. Urinary Studies in Children, 2-10 Years Old, Resident in a Hypo- and Hyperendemic Area. *American Journal of Clinical Nutrition* 20 : 1369-1377.

Von Neumann, John and Oskar Morgenstern
1964 *Theory of Games and Economic Behavior.* New York: Wiley.

Watabe, Tadayo
1967 *Glutinous Rice in Thailand.* Kyoto: Center for Southeast Asian Studies, Kyoto University.

Wharton, Clifton R.
1969 *Subsistence Agriculture and Economic Development.* Chicago: Aldine.

World Weather Records, 1951-60
1961 Washington: United States Weather Bureau.

INDEX

PUBLICATIONS OF THE CENTER FOR SOUTHEAST ASIAN STUDIES

Special Report Series

Number 1: J. A. Niels Mulder. *Monks, Merit, and Motivation. An Exploratory Study of the Social Functions of Buddhism in Thailand in Processes of Guided Social Change.* 1969, [v], 43 pages. (*Out of Print*).

Number 1: J. A. Niels Mulder. *Monks, Merit, and Motivation: Buddhism and National Development in Thailand.* Second (revised and enlarged) edition. 1973, vii, 58 pages. (*Out of Print*).

Number 2: David H. de Queljoe. *A Preliminary Study of Malay/Indonesian Orthography.* 1969, [v], 91 pages. (*Out of Print*).

Number 3: Chan Ansuchote. *The 1969 General Elections in Thailand.* 1970, [v], 44 pages. (*Out of Print*).

Number 4: Clark D. Neher. *Rural Thai Government: The Politics of the Budgetary Process.* 1970, [v], 60 pages. (*Out of Print*).

Number 5: David H. de Queljoe. *A Preliminary Study of Some Phonetic Features of Petani, with Glossaries.* 1971, [xiii], 114 pages. (*Out of Print*).

Number 6: Richard L. Stone. *Philippine Urbanization: The Politics of Public and Private Property in Greater Manila.* 1973, vii, 149 pages. Bibliography, maps, index. (*Out of Print*).

Number 7: Carl H. Lande. With the assistance of Shirley Advincula, Augusto Ferreros, and James Frane. *Southern Tagalog Voting, 1946-1963. Political Behavior in a Philippine Region.* 1973, vii, 160 pages. Maps and index.*

Number 8: Herbert J. Rubin. *The Dynamics of Development in Rural Thailand.* 1974, x, 159 pages. Maps, bibliography.*

Number 9: Harry Aveling. *A Thematic History of Indonesian Poetry: 1920-1974.* 1974, x, 90 pages. Bibliography.*

Number 10: Frederick Wernstedt, Wilhelm Solheim II, Lee Sechrest, George Guthrie, and Leonard Casper. *Philippine Studies: Geography, Archaeology, Psychology, and Literature: Present Knowledge and Research Trends.* 1974, 113 pages. Annotated bibliography, index.*

Number 11: Howard M. Leichter. *Political Regime and Public Policy in the Philippines: A Comparison of Bacolod and Iloilo Cities.* 1975, vii, 165 pages. Maps, charts, bibliography, tables, index.*

Number 12: G. N. Appell, editor. Contributors: G. N. Appell, I. D. Black, D. E. Brown, Jay B. Crain, Robert McKinley, Peter Metcalf, and Vinson H. Sutlive, Jr. *Studies in Borneo Societies: Social Process and Anthropological Explanation.* 1976, vii, 158 pages. Maps, bibliography, tables, index.*

Number 13: Douglas E. Foley. *Philippine Rural Education: An Anthropological Perspective.* 1976, ix, 114 pages. Tables, bibliography, index.*

Number 14: John A. Lent, editor. Contributors: Nat J. Colletta, Cynthia H. Enloe, John A. Lent, Daniel Regan, Nancy L. Snider, and Alan B. Wilson. *Cultural Pluralism in Malaysia: Polity, Military, Mass Media, Education, Religion and Social Class.* 1977, x, 114 pages. Charts, bibliography, index.*

Occasional Papers

Number 1: David W. Dellinger (ed.). *Languages, Literature and Society: Working Papers from the 1973 Conference of American Council of Teachers of Uncommonly-Taught Asian Languages (ACTUAL).* 1974, 85 pages. (*Out of Print*).

Number 2: Ronald L. Krannich, Herbert J. Rubin, Pratya Vesarach, and Chakrapand Wongburanavart. *Urbanization in Thailand.* Center for Governmental Studies, Northern Illinois University. 1974, iv, 116 pages. (*Out of Print*).

Number 3: Gerald S. Maryanov. *The Condition of Southeast Asian Studies in the United States: 1972.* In cooperation with the Southeast Asian Regional Council, The Association for Asian Studies. 1974, vi, 68 pages. (*Out of Print*).

Number 4: Donn V. Hart. *An Annotated Bibliography of Philippine Bibliographies: 1965-1974.* 1974, [iv], 160 pages. Author and subject index.*

Number 5: Donn V. Hart. *Thailand: An Annotated Bibliography of Bibliographies.* 1977, 100 pages, index, table.*

* These publications are available *only* from the Cellar Book Shop, 18090 Wyoming, Detroit, Michigan 48221. No *Out of Print* copies are available from the Center.

OTHER CENTER PUBLICATIONS

Theses on Southeast Asia Presented at Northern Illinois University: 1960-1971. An Annotated Bibliography. The Center for Southeast Asian Studies, Northern Illinois University, DeKalb, Illinois 60115. 1972, 13 pages. Bibliographical Publication No. 1. (Available *only* from the Center for $1.50, prepaid).

A Guide to the Southeast Asia Collection. Swen Franklin Parson Library. Northern Illinois University, DeKalb, Illinois. The Center for Southeast Asian Studies. 1972, 12 pages. Bibliographical Publication No. 2. (Available *only* from the Center for $1.50, prepaid).

Tenth Anniversary Report: Center for Southeast Asian Studies, Northern Illinois University, DeKalb, Illinois. The Center for Southeast Asian Studies, 1974. iii, 20 pages. (*Out of Print*).

Lee S. Dutton. *Current Periodical Holdings of the Southeast Asia Collection of Northern Illinois University Libraries.* The Center for Southeast Asian Studies, 1974. 32 pages. Bibliographical Publication No. 3, (*Out of Print*).

_____. *A Checklist of Philippine Pamphlet Holdings of Northern Illinois University Libraries.* The Center for Southeast Asian Studies, 1975. [iv], 57 pages. Subject index. Bibliographical Publication No. 4, 1975. (*Out of Print*).

_____. *Newspapers in the Southeast Asia Collection of Northern Illinois University Libraries.* The Center for Southeast Asian Studies, 1976, vii, 14 pages. Bibliographical Publication No. 5. (*Out of Print*).

CENTER FOR SOUTHEAST ASIAN STUDIES

The Center for Southeast Asian Studies was established in 1963 at Northern Illinois University, DeKalb, an hour's drive west of Chicago. Current NIU enrollment is about 22,000 students, including nearly 4,000 graduate students.

The Center's academic program concentrates on Thailand, Indonesia, Malaysia, and the Philippines. A minor in Southeast Asian Studies is offered at the undergraduate level, while a Concentration is available for both M.A. and Ph.D. programs. The doctorate is given in the departments of History, Economics, Political Science, English, and Psychology.

Fellowships and graduate assistantships are available for qualified students. In the past the Center has received regularly FLAS (formerly NDFL) fellowships for the study of Thai and Indonesian, the two Southeast Asian languages taught at NIU. The Southeast Asia Collection in the new Founders Memorial Library consists of approximately 25,000 volumes, while periodical holdings for this region total over 1,000 titles. Additional research materials on Southeast Asia are readily available in collections in the Chicago area, e.g., The Newberry Library (Ayer's Filipiniana Collection), The Center for Research Libraries, etc. There is a growing collection of Southeast Asian materials in the Anthropology Museum and the Music Department has a large collection of Southeast Asian musical instruments. About 250 Southeast Asian students annually enrolled at the University. Each year outstanding specialists on Southeast Asia visit the campus to give talks on their disciplinary specialties.

For various brochures describing the Southeast Asia program, write: Dr. Donn V. Hart, Director, Center for Southeast Asian Studies, Northern Illinois University, DeKalb, Illinois 60115.

The faculty affiliated with the Center include JAMES A. BEAUDRY (Sociology, Overseas Chinese); RICHARD M. COOLER (Art History, Thailand and Burma); DAVID H. DE QUELJOE (Foreign Languages, Linguistics, Indonesia, Malaysia, and the Philippines); LEE S. DUTTON (Southeast Asia Librarian); KUO-HUANG HAN (Ethnomusicology, Indonesia and Thailand); DONN V. HART (Anthropology, Philippines); JOHN F. HARTMANN (Foreign Languages, Linguistics, Thailand); CLARK D. NEHER (Political Science, Thailand), RONALD PROVENCHER (Anthropology, Malaysia and Indonesia); M. LADD THOMAS (Political Science, Thailand and the Philippines); HERBERT J. RUBIN (Sociology, Thailand); HAROLD E. SMITH (Sociology, Thailand), and CONSTANCE M. WILSON (History, Thailand).